Skylarks with Rosie

A Somerset Spring

Stephen Moss

Saraband

Published by Saraband
Digital World Centre, 1 Lowry Plaza,
The Quays, Salford, M50 3UB

www.saraband.net

ISBN: 9781913393045
ebook: 9781913393052
Printed and bound in Great Britain by Clays Ltd, Elcograf S.p.A.

1 2 3 4 5 6 7 8 9 10

Contents

To my lockdown companions:
Suzanne, Charlie, George, Daisy, David,
Kate and Mark; to James, six thousand miles
away in Japan; and, finally, to Rosie, whose
love for us all knows no bounds.

There are two ways of acquiring wisdom.
One – they say – is travelling far and wide.
The other is to stay in a location,
focusing ears and thought and eyes

on all that surrounds you in the one place
in which you choose (or are forced) to bide,
noting how the seasons slide
into each other,
the rise and fall of wind or cloud or tide
taking account of changes
and allowing them to guide
the path on which you step and stride.

Someday, though my friends would all deny it
(indeed, it would be to their great surprise),
I'll have circled all the tracks around this township
and discover I am well and truly wise.

Donald S Murray
From *The Man Who Talks to Birds* (Saraband, 2020)

Prologue

Our world is cribbed, confined and bound in as never before. Yet amidst all the fear and horror, there is one small but significant silver lining, as we reconnect with nature on our doorstep.

My Somerset garden is awash with birdsong: chiffchaffs, wrens, robins and a new arrival, the blackcap, all competing to see who can shout the loudest as spring gathers pace. Overhead, buzzards soar and ravens tumble, as delighted as I am to herald the new season.

But to hear a bird whose song is the definitive sound of the countryside, I must take my daily exercise: a walk with our fox-red labrador, Rosie, around Blackford Moor, the little patch of land behind our home.

I've seen some memorable birds here over the years. But as with any local patch, it's all about the commonplace; and here, and especially now, the ubiquitous bird is the skylark.

Skylarks with Rosie

A plump, triangular-winged shape rises up from the lane in front of us, then rapidly gains height, while continuing to deliver an outpouring of song: a rapid jumble of notes that seems to go on forever, even when the bird vanishes into the ether. As the dog and I stand and watch, I feel a new book coming on: *Skylarks with Rosie*.

*   *   *

That is how this little book began: in the whirling heart of the greatest peacetime crisis of our – or indeed any other – lifetimes. At first, things didn't really seem very different from usual, especially here in the Somerset countryside. But as the days went on, and the world regressed into what began to feel like the seemingly endless Sunday afternoons of my 1960s childhood, something unexpected happened: the coming of spring.

I say 'unexpected', yet those of us who feel a deep affinity with nature had been waiting for spring ever since New Year's Day. But two things were different this year. First, those of us *already* connected with the natural world were made to experience this seasonal rush right on our own doorsteps, rather than further afield; and second, the vast majority of Britons – those who until then had hardly been aware of the changing of the seasons – were connecting with nature, too.

Much in the manner of lifelong atheists experiencing a Damascene conversion to God and religion, the nation opened its eyes and ears to what had been going on every spring for their entire lifetimes, yet which until that moment they had been too busy, too preoccupied or simply too blinkered to notice. Birds were singing to defend their territories and win a mate – as they always do at this time of year – but now things had changed: they were not just being heard, but being listened to, in a collective human awakening to the joys of the natural world.

What I – and my fellow naturalists – have always loved about our lifelong passion is that engaging with wildlife, at any level, makes us feel better. A few years ago, scientists at the University of Surrey proved this link between nature and well-being, specifically with regard to birdsong. Those of us 'in the know' had, of course, already appreciated that nature was good for us; nevertheless, it was nice to hear it given official approval.

And now, to our huge surprise, we were sharing this love and appreciation for birdsong with the entire nation. For whether you lived in the city, as I did for the first half of my life, or the countryside, as I do now, you could not help but notice the daily rise in the volume, intensity and variety of birdsong, as spring took hold across Britain in 2020.

So, I hope this is more than simply a diary; though it does indeed trace my day-to-day experiences of this unique and unrepeatable spring, for three intense and memorable months from the March equinox to the June solstice. It is, above all, a record of how the nation fell in love with nature at a time of existential crisis; and how nature, without ever realising it, helped us get through to the other side.

Introduction

The Loop

Sunday 22nd March, my wife Suzanne's birthday, dawned bright, sunny and warm, so after I and my three teenage children had given her presents and sung the ritual song, she and I went for a walk. As always, we went round what we have come to call 'the loop': a three-mile route out and back from where we live, on the northern edge of the Somerset Levels. Shaped more or less like a letter Q, it begins with a stroll down the lane, followed by four sides of a rough square, then back up the lane to our home.

Our dog, Rosie, was with us, running gleefully ahead before stopping, turning and checking that we were still there. This is by far her favourite walk; and indeed ours, for although this flat, open 'moor' is not conventionally scenic, it does provide a series of views away from our home towards the high points of the Mendips to the north, the Wells telecommunications mast to the east and Brent Knoll to the west. Finally, as we return, we glimpse the reassuring presence of our village church tower to the south, guiding us back home.

Soon after we took the first turning on the square, a low-flying raptor appeared, heading away from us. Immediately both Suzanne and I realised that this was not a buzzard: the dark plumage, long, raised wings and determined flight marked it out as a female marsh harrier. I presumed that this was not one of the birds that breed down on the Avalon Marshes a few miles to the south, but a migrant, heading north-east on what may be the final leg of its journey from north-west Africa, where some of our birds spend the winter.

A good start, soon followed by a nicely varied collection of birds: a late fieldfare, also heading north; a chiffchaff – the first I had heard here this spring – endlessly repeating the two syllables of its name; and a female kestrel, hovering on wimpling wings in the fresh breeze.

We also came across two species that would have been unknown here even twenty years ago. The first was a Persil-white little egret, feeding in one of the deep-sided rhynes (the network of drainage ditches that criss-cross the whole of the Somerset Levels). The second was the exact opposite, colour-wise: two ravens flying overhead, uttering their characteristic 'cronking' call. More about both these charismatic birds later.

Returning home, an hour after we set out, we felt that self-satisfied glow of having done our daily exercise, walked the dog and spent this part of the weekend as it should be spent: outdoors, in each other's company – something we rarely get the chance to do in this increasingly frantic world. Little did we know that, within just twenty-four hours, all this was about to change.

*　　*　　*

Normal life for us, as for so many people, is always just that little bit too busy. Weeks are filled with the usual routine: work, for me mostly at home but with weekly trips to the university where I teach; and for Suzanne, the peripatetic life of a health visitor, helping new mums and their infants in the community.

The days begin early, when we wake at 6.30 to get the children ready for school, and end late, as we finish the chores before flopping down in front of the television. Weekends, which should be a time of rest and recuperation, are even busier, as we catch up on washing and ironing, cooking and cleaning, and ferrying the children to distant parts of the county to play rugby and football. There seems to be no time to stop and stare; nor, indeed, to stop and do anything.

Things did begin to change back in January when, after forty years in the NHS, Suzanne retired. I say 'retired', but with her active approach to life, coupled with a desire to do something different, she had already planned a new and exciting venture involving health, well-being and nature. When the crisis struck, however, she was still in the planning stage, and so we were enjoying the many benefits of her being at home all day long.

Soon afterwards, the children were home too, as school

was shut until further notice. Their response to this unprecedented situation reflected their very different personalities:
sixteen-year-old Charlie, who had been studying surprisingly hard for his GCSEs, was understandably delighted
that the exams had been cancelled; George, aged fifteen,
was relieved that school was over, as he prefers studying at
home; while Daisy, his twin sister, was temporarily hysteri-
cal at the thought of being cooped up with her parents and
brothers for what seemed like a lifetime. I could sympathise,
but as I pointed out, we all had to adapt. She and I, shall we
say, agreed to differ.

*   *   *

I work at home much of the time anyway, mostly in the
office in our garden, and so am used to a partly solitary life.
Nevertheless, because I have what is now called a portfo-
lio career – combining writing, teaching, leading bird tours
and giving talks – I do get out and about from time to time.
Now all that was over, at least for a while.

In late April 2020, at the very epicentre of the lockdown period, I would be turning sixty. It felt like a bigger
milestone than previous 'big birthdays', perhaps because it
sounds so ridiculously old. At least having teenage children
keeps me feeling younger than many of my contemporaries, many of whom who are embracing retirement with
what, to me, seems like excessive enthusiasm.

Last New Year's Eve, a time of year when I usually reflect
on how I might improve my hectic work/life situation, I
turned to Suzanne and told her that, as I would be reaching

that special milestone this year, I would start to slow down a little.

To be fair, she did her best to conceal her scepticism, but I could see that she did not really believe me. She had a point. I had already taken on an extra day a week teaching at the university. I had one book about to come out and a tight deadline for the next one. Two days later, I would travel to South Africa to observe swallows at their million-strong roost; and I was already committed to a number of talks and birding tours in the coming year, as well as trips to Japan, Turkey and Australia. In what way was I planning on slowing down?

And then, on Monday 23rd March, came the lockdown. First, the talks were cancelled, then the local bird tours, then the foreign trips, then Birdfair – the annual event that brings together bird and wildlife enthusiasts from all over the world. Domestically, school was closed, the children's sporting seasons brought to an abrupt end, Suzanne's new project put on hold, and any kind of social life – including my own long-planned birthday party – postponed for goodness knows how long.

My family and I – and the majority of people here in Britain and around the world – were on a kind of indefinite leave of absence from our usual lives. My work did go on – I could at least write and teach (online) when self-isolating – but everything else was turned upside-down for what felt like an indefinite period.

* * *

After the initial shock, we all fell into a comfortable – and rather comforting – routine. Suzanne and I got up a little later than usual, and after breakfast I would usually take my daily exercise by cycling around the loop. I have been doing this for almost a decade now, but I had only started keeping a proper list of the birds I saw there the previous January.

Before the lockdown began, I had already tallied almost fifty species on what I was beginning to regard as my 'home patch', compared with just over sixty during the whole of the previous year. I like a challenge and so, as the lockdown commenced, I focused on methodically recording what I saw and heard, either during the early-morning ride or on a more leisurely afternoon walk with Rosie.

At the same time, of course, I could hardly ignore what was happening in our garden. Spring is always the most exciting time of year here, as it is anywhere, but in recent years I have often been away at the critical time, missing the one or two weeks of the year when everything kicks off.

Being here every single day reminded me of when we first moved to Somerset back in 2006. For the first few years – until I found other, more distant locations – my garden was where I did the vast majority of my bird- and wildlife-watching.

The children were then just toddlers, and so inevitably we were confined to home and garden for much of the time. I documented our first full calendar year here in my book *A Sky Full of Starlings*, and a few years later extended my scope to the whole of our parish with *Wild Hares and Hummingbirds* (the latter a reference to the hummingbird hawkmoth, not its avian namesake). It now struck me as ironic that I had deliberately chosen to confine myself to my parish for those two books; with the lockdown, I was being forced to do so.

I also realised that, during the intervening decade or so, I had begun to neglect what was immediately on my door-step, or at least take it for granted. Instead of sitting quietly outside, waiting to see what flew overhead, or listening out for birdsong from the surrounding hedges, shrubs and trees, I would spend any spare time I did have at my two 'local patches': one a hidden corner of the Avalon Marshes, the other where three rivers meet – the Huntspill, Parrett and Brue – at what passes for coastline here in Somerset.

I had totted up an impressive total of birds at each of these locations: a round one hundred species at the inland patch, including bearded tit, cattle and great white egrets and ten species of warbler; and over one hundred and twen-ty-five species along the coast, the highlights being grey phalarope, Mediterranean gull and my best ever discovery, a splendid male red-backed shrike, perched alongside a flock of linnets on a sunny July morning.

All these made my garden sightings feel somehow less significant in comparison. Yet as it dawned on me that my garden and the moor behind my home were now the boundaries of my life for the foreseeable future, I again realised what I have always known: that birding isn't about the rare and unusual – exciting though they are – but the reassuringly regular and commonplace. In any case, this was what I would be seeing and hearing during this particular spring, and so, I decided, I had better start to enjoy it.

Week One

23rd–29th March

Blackcap

Work has carried on this week, more or less as normal. But one morning, even as I took the short walk from my home to my office – a distance of about fifty feet – I heard the sound of a blackcap.

This is the bird that John Clare called the 'March Nightingale', though to be honest comparing its sound to that of our greatest songster might be pushing it a bit. Nevertheless, the sweet song of this large, grey warbler was a welcome sign that spring is, little by little, arriving back on our shores.

Blackcaps – which sound rather like a cheerful robin or speeded-up blackbird – are normally the second migrant to return, after the chiffchaff, which started singing in the garden a week or so ago. This was the earliest singing blackcap I have heard in the fourteen springs since we first moved here. My excitement at this was tempered, however, by the thought that these early-arriving migrants are only here because of the changes wrought by the climate crisis; none of which will benefit them in the longer term.

Like the chiffchaff, British blackcaps don't have all that far to come. In contrast to later arrivals such as the willow warbler, these two species do not travel all the way from equatorial Africa, but only from the Iberian peninsula and north-west Africa, where they spend the winter.

I was delighted to hear the blackcap's sweet and tuneful song, which will continue for the next couple of months, as a pair are nesting somewhere deep inside the tangled scrub of brambles, hawthorn bushes and cider apple trees at the bottom of our garden.

Other signs of spring that morning included singing dunnocks, robins and wrens, flocks of goldfinches twittering overhead, their yellow wing-flashes catching the sun, and a local scarcity – a mistle thrush – whose bulky shape I caught sight of as it flew up into one of the cider apple trees. Despite what appears to be ideal habitat – large, open fields with tall trees where they can sing and nest – I rarely see mistle thrushes around here. Another reason to welcome the chance to stay put: to get a proper picture of which birds, butterflies and other species actually use our garden.

That evening, on the BBC News, we saw another side of the Prime Minister, who seemed to have foregone his usual rather irritating jocularity, and adopted an uncharacteristically serious tone, as he announced the beginning of lockdown for the whole country. A week or so ago, he appeared like the figure of Death in that Ingmar Bergman movie, intoning these fateful words: 'Many more families are going to lose loved ones before their time.' Not a very pleasant thought.

*   *   *

Fortunately for our collective sanity, health and well-being, the start of lockdown has seen a welcome shift in the weather across much of the UK. February had been the wettest on record, and March was not much better – staying cool, cloudy and damp for most of the month, making our back lawn squelch underfoot.

But then the weather systems turned from low to high pressure, and although the mornings have been chilly, the skies are clear, the sun is shining and (so long as you ignore the TV news bulletins) all appears well.

I can't decide if the weather gods are mocking us or being kind. And so it is with the authorities, who encourage us to take our daily exercise while simultaneously issuing grim warnings against travelling to beaches and beauty spots. I find myself becoming increasingly frustrated at the dithering and lack of clarity in the government's response to this crisis. Our Prime Minister constantly evokes Winston Churchill but is like a parody of a proper leader – all bluster and no substance. (At the time, of course, we failed to realise that his more than usually dishevelled appearance was because he too was suffering from the early effects of this potentially deadly virus.)

Skylarks with Rosie

*　　*　　*

Like many others not stricken by the disease, I don't think I have ever felt so fit and healthy, or cycled so much. Meanwhile, Rosie has never been on so many walks: usually two a day, either with me or Suzanne, or one of the children. They are enjoying the fine weather, too, and are escaping the confines of their bedrooms for unaccustomed doses of fresh air.

Charlie has taken up the guitar again, after learning it until he was eleven and then giving up. He has a natural talent (unlike me, who, having learned the piano for seven years as a child, cannot play a note) and is serenading us with electric and acoustic versions of Deep Purple's *Smoke on the Water*, Nirvana's *Smells Like Teen Spirit* and various Beatles songs, including, rather appropriately, *Blackbird*.

Wildlife-wise, the last week of March always finds me in a state of flux: delicious anticipation mixed with frustration. The anticipation comes because the onrush of spring – that flood of new emergences and arrivals from April onwards – is just about to start; the frustration because, like Diana Ross, I'm still waiting.

On the weekend before the lockdown, the weather switched from gloom, rain and cold to warmth and sunshine; and with it came the first butterfly of the year: a peacock flitting along the cider apple hedgerow next to our garden, in a stiff easterly breeze.

Peacocks are the Rocky Balboa of the butterfly world: large and chunky, with a powerful flight and combative nature – often chasing off any rivals that dare to venture into their territory. Their name comes from the prominent false 'eyes' on the leading edge of their forewing, which serve two purposes. First, they startle any avian predator; and second, if that fails to work, the bird will peck at the false eye rather than the butterfly's body, thus enabling it to escape, albeit with its wings a little frayed at the edges.

Three days later, on 24th March, the sun really began to warm up, and two other classic early spring butterflies joined the peacock on the wing: the comma and brimstone. This trio, along with the small tortoiseshell, spend the winter as adults, often holed up in the cobwebby corners of our garages, sheds and other outbuildings, to emerge from their 'hibernation' on a warm, sunny spring day. I have seen them as early as January, but this year they appeared quite late, thanks to the unseasonably cool and damp weather, which thankfully appeared to be at an end.

These welcome signs of spring coincided with some equally welcome news from my eldest son, David. He and his partner, Kate, are prematurely ending their world tour and flying back from Japan to move in with us – swapping the delights of Osaka, Kathmandu, Istanbul and Bogotá for our sleepy Somerset village. I do feel sorry for them having to put their travel plans on hold, but I am also secretly looking forward to having them home. At times like these, you want your family as close as possible (though, ironically, David's younger brother, James, lives permanently in Japan and will be staying put for the duration – however long that might be).

*   *   *

The day before David and Kate's return, I noticed a quartet of ravens flying high over our home, uttering their deep, basso-profundo calls, twisting and turning in the clear blue sky and looking as if they were having great fun. It seemed a bit soon to be a family party of adults and offspring (though ravens are amongst our earliest birds to start nesting); and, indeed, after posting my sighting on the Somerset Birding website, I was informed that because ravens do not usually breed until they are at least three years old, these were almost certainly young birds – the corvid equivalent of rowdy adolescents – simply letting off steam.

A few days later, I came across a flock of more than thirty ravens in the far corner field: attracted by the carcass of a new-born calf that had died overnight. They hopped about in the tussocky grass like animatronic creations, their glossy blue-black plumage glinting in the midday sun. Every now and then, they rose up into the air for no apparent reason, and squabbled with one another, uttering their deep, throaty cries.

Seeing the ravens reminded me of a trip with my mother to Snowdonia back in the summer of 1975, when I saw ravens in the wild for the very first time. Who could have predicted that less than half a century later, this magnificent corvid – the largest of all the world's six thousand or more passerines (perching birds) – would now be found more or less all over the UK.

Even here in Somerset, their numbers have exploded during the time we have been living here. For the first six months or so after we moved, I didn't see a raven at all; and it took almost another year before one finally appeared close to home, in a nearby apple orchard. This spring, for the first time, they have been more common than carrion crows, rooks and jackdaws.

Like many other all-black birds, including the crow and cormorant (though not, oddly, the blackbird itself), ravens are generally viewed as an ill omen. I, for one, am always glad to see them, especially on a fine spring day when they perform their aerial acrobatics in the blue skies above. Besides, we have had our fill of ill omens recently.

* * *

At the end of this first lockdown week I took one of my regular early-morning cycle rides around the loop and noticed a rise in the volume and intensity of birdsong. Wrens and skylarks still dominated: the former tucked inside the thick hawthorn hedges; the latter living up to their name by hanging like dust-spots in the sky.

Despite their diametrically opposed lifestyles, the wren and skylark do have one thing in common: they both sing with such incredible force and exuberance. The tiny wren produces short, intense bursts, full of bravado and brio; the skylark sings at a more leisurely pace, yet with a persistence that seems beyond imagination. I defy anyone listening to these birds not to feel as if the cares of the world have been lifted – even if only temporarily – from their shoulders.

I also saw and heard my first reed bunting here this year: a male sporting his splendid black hood, looking like a medieval executioner. This is a bird I always enjoy coming across, and once again I was instantly taken back in time, to almost exactly half a century ago. I was at Shepperton gravel pits, a few hundred yards from my childhood home, and it was the very first time I had ever come across this species. Somewhere in our voluminous collection of storage boxes, which we still need to sort through (another job for lockdown?), I have a pencil drawing I did at the time: clearly showing the male bunting in all his glory. It was the last time I ever tried to draw a bird.

Yet another sighting that took me straight back to my childhood was a little egret, fishing methodically in one of the watery rhynes that criss-cross the moor. The same year that I drew that reed bunting, we went for a family holiday to the Hampshire coast. One day, as a special treat, we headed further west, into Dorset, on a visit to Brownsea Island.

I can still remember walking into one of the hides overlooking the lagoon, lifting the shutters, and being confronted with a vision of purest white, perched in a tree on the opposite side of the water. My first little egret – and, indeed, my first proper 'rarity'. Apart from occasional sightings around the Mediterranean later on, this was the only one I saw in Britain until 1989, when the invasion occurred that would lead to this small and elegant heron becoming a common and familiar waterbird. Yet even now, perhaps because of the thrill of that very first childhood sighting, I can never quite accept little egrets as part of our landscape; for me, they will always retain a whiff of the exotic.

*　　*　　*

Week One

Already, just a week into lockdown, the newspapers, TV and radio have all been reporting a strange and unexpected phenomenon: the sudden realisation that nature is all around us. Questions have been asked in both conventional and social media: is this a particularly unusual spring, during which wild creatures are somehow appearing more often, and in greater numbers, than ever before? Or are we simply noticing it more than usual, because of the very nature of lockdown?

In fact, both suggestions contain more than a grain of truth: the appearance of feral goats in the deserted streets of Llandudno, and fallow deer in the East London suburbs, are clearly a result of the lack of people and motor vehicles in the streets.

People are also reporting sightings of normally shy mammals such as stoats, weasels and even moles – a creature that, despite its ubiquity, I have still never seen alive. On the continent, where lockdowns began earlier than here, dolphins have allegedly returned to Venice's network of canals, whose waters turned an unfamiliar shade of blue. (It later turned out that the dolphins had actually been filmed near the port of Cagliari, in Sardinia. Pity.)

The lack of pollution brought about by this unprecedented downturn in social and economic activity doesn't just bring benefits to nature, but also to us: indeed, it is possible that the number of deaths from Covid-19 will be balanced by the reduction in those due to air pollution; the problem is that, first, those are much harder to measure, and second, any death is a tragedy, so these kinds of comparisons do seem invidious.

The sudden surge in awareness of birdsong has clearly been down to three very human factors: the lack of extraneous noise from aircraft and road traffic; the fine, sunny weather; and, of course, the fact that, in many cases for the very first time, people have been taking a break from their busy lives, and so have had time to listen.

(Interestingly, the Radio 4 programme *More or Less* later reported that birds might actually be singing at *lower* volumes than usual. This, a scientist noted, was an example of the 'Lombard Effect', in which human beings tend to speak louder when background volumes increase; contrariwise, we speak more quietly when there is little or no extraneous noise. Did the same apply to birdsong? It appears that it might: measures taken of a singing chiffchaff near Manchester Airport suggested that the bird almost doubled its volume when a plane flew overhead. Without aircraft noise, they no longer needed to sing quite so forcefully. Even so, our *perception* was clearly that birds were singing more loudly – even if they were not.)

There seems to be an insatiable desire in the media to make sense of this phenomenon. On the last Saturday of March, I was invited onto the *Today* programme's regular 'Nature Notes' slot, where Martha Kearney quizzed me about the way birdsong was now flooding across the nation like some kind of soothing balm.

Even as we listened to three of our commonest songbirds – the blackbird, robin and wren – I could feel the way birdsong comforts us. However, I did go on to explain that they are not singing for our benefit, but to repel rival males and attract a mate, or what the poet A.F. Harrold famously summed up as 'fuck off or fuck me'. I didn't mention that particular phrase on the radio, though.

Afterwards, I wondered if this new fascination with nature would last, or would the media – and the great British public – tire of the novelty of twittering robins, trilling wrens and fluting blackbirds, and go back to tweeting of a more human kind?

Week Two
30th March–5th April

Swallow

Before I continue, let me take you on a guided tour of the loop.

I go out of our gate and turn right, heading north down the lane – the only (very shallow) gradient on the whole of the three-mile route. At the bottom of the lane, by the hamlet of Perry – where presumably pear cider used to be made – I turn sharp right over a bridge across a shallow rhyne, then out into a patchwork of open fields, mainly used to graze sheep or produce silage (or, during the last two warm, fine and sunny summers, hay).

At various points, signs tell me that *This road is liable to subsidence*. This is a rather superfluous warning, given that it applies to virtually every lane around the Somerset Levels, whose peaty soils are prone to movement, making driving along them feel like a ride on a particularly bumpy rollercoaster.

The hedgerows bordering these fields are thick and broad – sometimes three or even four metres across – and consist mainly of hawthorn and blackthorn, with a few other native plants interspersed among them. The hedges vary considerably in height and are dotted with taller trees; many are also above water-filled ditches or rhynes, which produce a ready supply of insects, on which the half-dozen warbler species that breed here feed their young. Were the hedgerows tidier, and the water absent, I suspect that this would be a much-depleted area, bird-wise.

After half a mile, I enter the loop itself, with the option of turning right (as Rosie invariably does) or going straight on towards the neighbouring village of Chapel Allerton. I can see the Mendip Hills in the distance, with the landmarks of Crook Peak and Cheddar Gorge. The centre of this square is criss-crossed by a network of rhynes, going either west to east or north to south. Some are clear, others clogged up with reeds and duckweed.

As the road bends round to the right, then right again, the view changes subtly; the eastern lane has taller hedges and trees, while the southern one is completely open, with no hedgerows on either side. This takes me back to the first junction, from where I head back the way I came to our home.

Every season – indeed, every visit – is different here. I recall some memorable moments: a flash of orange revealing a female redstart one May evening, as I drove George to football training; a fall of whinchats and wheatears on a

September morning, just a year after we moved here; and, best of all, unforgettable views of a short-eared owl on a snow-blanketed Christmas Eve a decade ago, when this bird, with its staring yellow eyes, was the only living creature we came across on the entire walk.

Now, I am seeing the whole place in a new light: not, as some people perhaps understandably feel about their lockdown location, as a prison, but as my glimpse of freedom on my daily walk or ride.

*   *   *

Each day this week, especially now that the clocks have gone forward, I have been feeling that familiar sense of expectancy – as I always do when March gives way to April. Things are moving on apace, as spring suddenly begins to accelerate, like a cyclist freewheeling down a steep hill.

The next few weeks will see a tsunami of activity in the natural world: more butterflies, more bumblebees, and even more birdsong, as millions more migrants return to our shores. Yet I still find it hard to believe that my little corner of the English countryside will soon be home to birds that have spent the last six months flying around the African savannah, in the company of lions, elephants and giraffes.

For the whole week I have been on tenterhooks: awaiting the bird I've been writing a book about for the past year – the swallow. In between feverishly scanning the skies above the garden, I also checked the Somerset Birding website and Twitter to see if swallows had been sighted yet, elsewhere in the county or beyond. It appears they have not, and the waiting goes on.

Besides, winter has not quite left us yet: the mornings are still close to freezing, as high pressure sits stubbornly above us, bringing light northerly winds. One morning, I came across a loose flock of a dozen or so thrush-sized birds, flitting rather clumsily along the tops of the hawthorn hedge on the eastern side of the moor. They were fieldfares, passing through the parish on their way north and east to their breeding grounds in Scandinavia or Eastern Europe.

Well over a million fieldfares spend the winter in Britain, and during November and December they can be the commonest bird on the moor, feeding on hawthorn berries or digging in the muddy fields for worms. Amidst all the springtime breeding activity, it felt odd to encounter these bulky thrushes, who looked like partygoers who have outstayed their welcome and were now trying to sneak away, unseen. They certainly seemed out of place as the volume and intensity of birdsong continued to rise.

Linnets sang their scratchy yet tuneful song from the top of the same hedgerows, the males showing off their splendid pink breeding plumage. In Victorian and Edwardian times, linnets were popular cagebirds, and were later featured in the 1919 music-hall ditty *Don't Dilly Dally on the Way*, which I remember my nan singing to me, as she sat at the piano in our tiny front room. The lines that always stuck in my mind were – in my nan's version – 'Off went the van with me old man in it / I followed on with me old cock linnet'. Funny how, at this time of confinement and reflection, almost every bird I see sparks childhood memories – some pleasurable, others rather poignant and sad.

*   *   *

On Saturday 4th April, at last, it happened. The wind moved round to the south, the air warmed up to a very pleasant temperature, and I rummaged in my wardrobe for my shorts, sandals and a summer shirt. As I sat outside enjoying breakfast and reading the newspapers, my first swallow of the spring cruised effortlessly overhead.

I find it hard to put into words how I felt about seeing this lone bird. Having spent much of the past year writing a biography of the swallow, and watched them both here in Somerset and at their winter roost in South Africa, I have come to know and love these birds with an intensity that has surprised even me. Suffice to say there was a tear or two in the corner of my eye as I watched this little bird head off towards the north.

The arrival of the swallow – along with those balmy southerly winds – opened up the floodgates. Later that same day, Suzanne and I were sitting out on the lawn when a white butterfly with Jaffa-coloured wingtips fluttered past: our first orange-tip of the spring. Like swallows, these attractive butterflies seem to mark the tipping point when winter really is over and summer is just around the corner: that association, and their delicate markings, always fill me with joy when I see them.

Once again, this brought back two vivid memories: not from my own childhood, but those of my youngest and oldest sons. In 2011, when George was just six, he came back from school one day and casually announced that he had seen 'one of them white butterflies with orange on its wings'. I gently explained to him that he could not have seen an orange-tip, as it was only 21st March, and they do not usually emerge until the first week of April. The next day I had to eat my words, when I discovered that there had been a major hatching of these cracking little butterflies all over southern Britain. I knew I shouldn't have doubted him.

Then, like a Russian doll emerging from inside another, an even more distant orange-tip memory floated into my mind. Thirty years ago this spring, I was sitting in my mother and grandmother's garden, when I noticed a strange butterfly I couldn't recall ever seeing before. Bear in mind that, at the time, I was singularly obsessed with birds and could only identify a handful of butterflies.

David – then aged three – quietly went off into the house and emerged with a book on British butterflies that my mother kept by the kitchen window. With all the confidence of early childhood, he opened the book and pointed to a picture of the very creature that was flitting from flower to flower: an orange-tip. I find great comfort in these odd links between past and present, linking the two children together via a single butterfly.

*　　*　　*

The nation, meanwhile, is agog with the news that the Prime Minister has been rushed into hospital after failing to recover from what we were persistently told was only a mild case of Covid-19.

Like most people, I wish him a speedy recovery. But I also

can't help recalling the way he behaved at the beginning of the crisis: not just shaking the hands of Covid-19 patients, but boasting about it; failing to get a grip on imposing a lockdown, in some kind of misguided belief in freedom and liberty; and, most of all, continuing to score cheap political points instead of genuinely uniting the nation as his hero Churchill would surely have done.

So, I hope that if and when he does recover, he will emerge with a little more humility than before. I hope that will happen, but do not really believe it.

Week Three
6th–12th April

Buzzard

As Easter approached, the fine and sunny weather continued, though the persistent easterly winds remind me that
it is still April, not June. I read somewhere that, for many
people, a sense of time has vanished: there are only three
landmarks in their day – coffee o'clock, tea o'clock and
wine (or beer) o'clock, at 11am, 3pm and 6pm respectively.

We have discovered a variation on this theme: 'buzzard
o'clock'. This is the hour or so from 11am to noon when,
having done a couple of hours' work, I emerge from my
garden office, join Suzanne and Rosie (the children are
normally not up yet) and scan the skies for raptors, as we
drink our coffee.

Again, I recall as a child the sheer excitement of seeing
a buzzard for the first time: on a trip with my mother to

North Wales. Twenty-five years later, I spent a winter week in the Netherlands to make a film with Bill Oddie, and we remarked on how odd it was to see so many buzzards, especially in a marshy, lowland area not all that different from where I live now. Yet nowadays I am surprised if I do *not* see a buzzard, wherever I walk, cycle or drive.

As if launched by some unseen signal around the time of elevenses each morning, buzzards begin to appear in the clear blue skies, which are now mercifully uncontaminated by aircraft and their vapour trails. We often hear them just before we see them: that rather wimpy mewing sound, like a petulant cat, followed by the appearance of the bird itself, rising rapidly on thermal currents of warm air, which take it from just above tree height to way up in the sky in a matter of a minute or two.

Suzanne has much better eyesight than me – helpful for spotting distant raptors. One morning, she drew my attention to a movement in the far west and, amongst a small group of gulls, there were half a dozen buzzards, soaring on broad, open wings. This became a daily occurrence, until by Good Friday there were no fewer than eight of them flying together.

We also saw a pair going down into a tree a couple of properties along from us; they are almost certainly nesting there. From time to time, the male rises up then folds his wings closed and plummets down like a ski-jumper, before opening his wings at the very last moment and shooting back up into the air. Occasionally, the female deigns to join him.

On one occasion, another raptor appeared, this time speeding eastwards at mid-altitude: a peregrine, the first seen over the garden for twelve years, and only the third I have ever seen here. It always amazes me that this bird, which usually appears to be in cruise control, is the fastest creature on the planet – their flight seems so leisurely when they are not hunting.

*   *   *

Good Friday – and every other day of this Easter weekend – dawned dry and sunny, and with the promise of some real warmth; more like late May than early April. Once again, it is as if the weather is mocking the nation for wanting to enjoy the Bank Holiday weekend. I remember Easter two years ago – admittedly at the end of March (why can't they simply fix a mid-April date and let us get on with it?) – when it was so cold it felt like February. Now, we were basking (or at least walking briskly) under cloudless skies.

The night before, we welcomed David and Kate, who had been under self-imposed lockdown. It was good to hug them at last.

On the butterfly front, I have been struck by the presence of two species: the comma and the holly blue. The former I do sometimes see in the garden, but the latter is barely an annual visitor. The first holly blue, on 9th April, was our earliest ever by a day; but after that I saw them almost every time I entered or left my office – a perfect vision of blue (a colour usually associated with summer butterflies) as they flutter along the hedgerow. The commas seem to have taken up residence in the 'messy bit' (actually, *very* messy bit) at the end of our garden, beyond the dog fence.

By Easter Saturday, no fewer than eight species of butterfly were on the wing – more than I have ever seen by this early stage in the season. That may not be just down to the fine weather, but also because I have been spending so much more time in the garden. The most recent to appear was the speckled wood, a pair of which flew in circles around one another, in a very public courtship display, as I mowed the lawn.

The one still missing, which I would have expected to see by now, is the small tortoiseshell – a species once so common we used to take it for granted, but which has suffered declines in recent years.

That night, as I was getting ready for bed, I heard our first owl of the year – a tawny – which came as a relief, as I thought they had disappeared from here after Suzanne found one dead by the side of the road a couple of years back. A few weeks later, on two successive afternoons, presumably the same owl hooted twice in broad daylight.

The same evening, I noticed the first bat, too: from its small size and fluttering flight action, presumably a common or soprano pipistrelle.

*   *   *

Earlier that day, I made my second appearance on the *Today* programme, this time talking about two returning migrants: the swallow and the cuckoo. I did mention the 'miracle of migration' (sometimes clichés just work) and also the fact that swallows – which weigh about the same as half a standard packet of crisps – often return to the very same place where they hatched out a year before. I got brownie points from my urban followers on Twitter by also talking about house martins – the town and city equivalent of the more rural swallow.

Later that day, I took part in my first public virtual event: an online literary festival. A young writer, Elli Wilson, did a remarkably professional job of encouraging at least sixty participants to take part in a controlled writing exercise, all via Zoom – a word that until now I had only associated with *Thunderbirds* and a type of ice lolly.

I read a piece in the *Guardian* pointing out that the lockdown is bringing unexpected benefits for people with a disability, in that it enables them to participate equally in events such as this – and to virtually 'visit' art exhibitions, theatres and music concerts, all of which are now online – in a way that they were often unable to do under what we used to think of as 'normal circumstances'. The unexpected benefits of this crisis may be outweighed by the downsides, but they should still be cherished when they do occur.

Week Four

13th–19th April

Mute Swan

Week four of lockdown, just after Easter, started with another change in the weather: this time from the warm and sunny conditions we have come to take for granted, to a very cool north-easterly wind, blowing right across the open moor. This turned my daily early-morning bike ride into a tale of two halves: the wind against me going out, and behind me coming home. I heard little or no birdsong on either leg.

I have been taking advantage of the now empty weekends in ways that I might perhaps have predicted. Last weekend, I tidied our sitting room and utility room, began working through the mess that has engulfed my garden office, and put my CDs in alphabetical order (even though last year I digitised them all and rarely listen to them now).

In a more public-spirited (and less anally retentive) task,

Week Four

I began tweeting daily extracts from the book I wrote with my friend and colleague Brett Westwood, *Wonderland*. Conveniently, this consists of three hundred and sixty-six daily entries on Britain's wildlife, half of them written by me and half by Brett. We joke, only semi-seriously, that you can always tell who wrote which entry: if you have heard of the plant or animal, it was me; if you haven't, it was Brett. Few if any people have quite such an encyclopaedic knowledge of Britain's wildlife as he does, and his fascinating entries on oil beetle, ivy-leaved toadflax and lemon slug prove the point.

One unexpected effect of the crisis – for me, at least – is that when I think I really should get in touch with someone, instead of either putting the thought to one side or sending an email or text, I am actually picking up the telephone and speaking with them directly.

So, I called Brett, partly to let him know I have been tweeting (though he does not indulge in social media himself), and partly just for the pleasure of hearing his voice. Sadly, he informed me that his mother, who I knew had been ill for a very long time, was now very close to death. I offered my condolences, having been through the same thing myself, before the conversation turned, as it always does, to nature, and we discussed the progress of spring migration.

Hanging up the phone, I thought of all the people having to deal with the crisis of a loved one's death – whether from Covid-19 or not – in these difficult times. The thought of not being with someone you love, to comfort them in their last hours, fills me with a profound sadness. Nature may be a great comfort, but there are some things it cannot resolve.

*   *   *

Later that day the wind dropped and – for the first time this spring – I heard the subtle but distinctive sound of a stock dove, coming from the tall ash tree at the bottom of our garden. Whereas wood pigeons have a five-note call (often transcribed as 'my toe *is* bleeding', with the stress on the middle note), and collared doves utter a three-note one (in Brett's view, chanting 'u-*ni*-ted', or for me, 'I'm *so* bored'), the stock dove makes do with just two: a long first note followed by an abrupt second one: 'Who-whoop!'

That's quite an extrovert sound for a bird that is surely the most overlooked in Britain. Stock doves look very like domestic or feral pigeons – plain grey, without the distinctive white collar and wing bars of the wood pigeon, or the pinkish-brown tinge and black neck-ring of the collared dove – and as a result they pass under the radar for most people.

Stock doves are not only widespread and common – especially here in rural Somerset – they are also quite a charismatic bird, at least for a member of the pigeon family. The name 'stock' actually comes from an Old English word for the lower tree-trunk left standing after the rest has been felled, where the doves often choose to nest. 'Stock' also fits the rather taut, compact shape of this dove rather well.

I think of stock doves as the Barry McGuigan of the bird world: they have a strangely pugilistic quality, punching above their weight, belied by their rather soft facial expression. In flight, they show a distinctive black trailing edge to their wings; on the ground, if you are lucky enough to get a close view of these rather wary birds, the custard-yellow base to the bill and the oily purple-and-green sheen of the neck patch are distinctive. As they often do at this time of year, the male and female were performing a beautifully synchronous aerial display over my garden, staying in perfect formation as they did so.

Seeing stock doves – along with the collared doves that perch on next door's roof – reminds me that one of their number, the turtle dove, is absent. The imminent extinction of the turtle dove is, like the feeling we have towards the people who have died in this terrible pandemic, a genuine bereavement. That is not to minimise the human loss we feel, but to emphasise that when a bird like the turtle dove simply disappears from our countryside, the sense of loss is a very real one.

* * *

On the Friday after Easter, I took part in another new online phenomenon. At the very start of the crisis, Chris Packham and his stepdaughter Megan McCubbin entered lockdown at their home on the edge of the New Forest. To connect with the wider natural history community, they came up with the 'Self-Isolating Bird Club', quickly attracting thousands of followers on Twitter and Facebook.

That morning, I had been invited to talk about nature writing – and about the way the lockdown has forced us to focus on nature in our local area. I began by suggesting people keep a nature diary of what they are seeing and hearing in their garden and on their daily walk or bike ride.

It turned out to be the only wet day for weeks, but the birds were still lustily singing in my garden during the whole of the broadcast.

Chris and Megan also gave a timely plug to a competition organised by my friend Lucy McRobert, who is encouraging the nation's children to write about 'Nature on Your Doorstep'. (A few weeks later she told me that delighted parents were contacting her in droves, some saying that their nature-obsessed children were finally spurred on to write about their passion; others that their writing-obsessed children had found nature enjoyable to write about. Just one of many examples of how the crisis was leading to an outburst of creativity, social bonding and kind-heartedness.)

What I have always loved about Chris is that if he tells you something, he really means it; there is no soft-soaping, or meaningless praise, for its own sake. So, I was very touched when, at the end of the event, he said some thoughtful words about me. I think we are all learning that, at a time of personal and national doubt about our future, the simple quality of kindness can never be overrated.

*   *   *

Over the past week or so I have been struck by the unseasonal gathering of as many as five hundred large gulls – split evenly between herring and lesser black-backed – in one of the fields along the western edge of the moor. By this time of year, I would expect these birds, virtually all of which are in full adult breeding plumage, to have formed pairs, and either headed westwards to the coast, where they breed on the island of Steep Holm in the Bristol Channel, or northwards to Bristol, Bath and Cardiff, where many now choose to breed on the roofs of city buildings.

A closer look revealed that the birds were indeed mostly in loose pairs – I even saw one pair of herring gulls mating briefly – but clearly the main reason they have come here is to eat. The farmer has spread liquid slurry onto the surface of the field, which must be attracting a wide range of insects and other invertebrates on which these huge birds – in the absence of herrings and other fish along our coasts – now mostly feed.

Gulls are not universally popular, either with the human traffic here (mainly dog-walkers and cyclists) or with the other wildlife. Crows and ravens frequently attack them, while one day, right behind my home, I noticed a fox being mobbed in turn by a gull, which soon made it flee into the hedgerow at the edge of a grassy field.

The mobbing of birds of prey often gives away their presence: usually crows attacking buzzards, but occasionally something more intriguing. One morning this week I was sitting outdoors when the sound of an angry crow drew my attention to a passing red kite – the first I have seen here for five years – drifting low over the ash trees at the bottom of the garden.

Later that day, a resident pair of raptors appeared: male and female sparrowhawks, the latter almost half as big again as her more compact mate. Suzanne and I watched enthralled as they began their courtship display: floating up into the blue sky, then twisting and turning before heading down towards the ash, where they have bred in the past. My fingers are well and truly crossed for a repeat performance this year.

When I was growing up, sparrowhawks were so rare – their numbers devastated by the use of chemicals such as DDT – that I saw barely more than a handful in my first two decades as a birder. Today, they have bounced back, and although they remain elusive, I do see them pretty regularly.

The migrants continue to arrive. The following day, George was sitting in the garden with Suzanne when they spotted a lone house martin overhead: that lovely, compact little bird that, as Bill Oddie once pointed out, looks like a miniature version of a killer whale.

His gentle twittering (the house martin's, not Bill's) was echoed by a resident species, the greenfinch, whose wheezes, trills and slow, deliberate display flight have been a regular sight this spring. Yet just three or four years ago, a whole year could go by without me hearing a single greenfinch – not just in the garden, but right across Somerset. This was because of a population crash caused by a fungal disease, which seemed particularly lethal to greenfinches. Fortunately, they appear to have acquired some immunity, and numbers have risen significantly. It's good to see them again.

*   *   *

The patience of a female mute swan as she sits on her eggs for six whole weeks, from the start of incubation to hatching, is nothing short of incredible. There is a Zen-like quality to her pose: just occasionally she leans over the edge of the nest and rearranges the reeds that form the majority of the structure.

The nest is conveniently placed halfway around the loop: a little way along the rhyne from a T-junction between two of the lanes. Suzanne and I first came across it on a joint cycle ride one weekend near the start of lockdown, and since then, every time I pass this point, I either stop to take a look, or simply glance across to make sure she is still present.

Today, the male had joined her, standing guard a few feet away. Not that I would fancy the chances of any predator trying to make a meal of her or her eggs: swans are notoriously feisty, and even though the commonly held notion that 'a swan can break a man's arm with a single blow of its wing' is utter nonsense, they are still able to ward off most attackers.

And so, we and the swans wait: for the red-letter day when the cygnets will hatch out, fluffy balls of grey for us to marvel at. While I was watching, I marvelled at the incredible song of a wren, whose nest is somewhere deep inside the hedgerow just a few yards away from that of the swans. One of our tiniest birds alongside our largest, yet both with the same, single-minded aim: to successfully raise a family.

*   *   *

Sunday 19th April was my eldest son David's thirty-third birthday. We gathered to celebrate and watch videos of his younger siblings that he – being more technically literate than me – had managed to salvage from a hard drive and put onto our television.

Not for the first time I was struck by the difference between my older and younger children. David and James (who is about to turn thirty) are the last of the analogue generation – those who can still remember life in a pre-digital world.

In stark contrast, Charlie, George and Daisy take for granted that they can access anything they want via the Internet, at any time of the day and night. And because we bought a video camera when Charlie was born, we have a visual record of the three of them growing up; something that David and James do not have. Suzanne and I barely even have any *photographs* of our childhood, let alone moving footage.

I wonder how this affects memory: after all, my twins are watching videos of themselves literally from the day they were born, as well as significant landmarks such as birthdays and Christmases in the years before they could possibly recall the events themselves. And, of course, they have their lives laid out on social media platforms. The art of writing about your distant past seems to me to rely on selective recollection of events – as Clive James called them, 'Unreliable Memoirs' – so what will people do when everything is not only available but there for others to see what really happened?

I suspected that we were not the only family catching up with videos from our past during these strange times: and was left both happy, and at the same time a little saddened, at seeing a past which is, apart from this rather wobbly footage, now out of reach.

Week Five

20th–26th April

Lesser Whitethroat

Having been a very casual Twitter user for the past few years, during lockdown I have taken to the social media platform like a very eager duck to water.

That's partly because there are so many interesting sightings being reported up and down the country; partly because I can share my own excitement at what I am seeing here on my home patch; but mainly, I think, because it allows us to connect with one another over our shared passion for nature at this difficult time. Never has the saying that 'Facebook is full of people you know and wish you didn't, and Twitter is full of people you don't know but wish you did', been more accurate.

Actually, that's not quite true: I have been exchanging tweets with old friends (including my longest-standing one, Glyn, who I have now known for more than half a century) and new ones, including 'flygirl' – Erica McAlister of the Natural History Museum – and 'Wenshine', my American friend Wendy Clark, whose partner Bill Thompson died in 2019, stupidly early, and with so much to give.

Then there are people I don't yet know but would like to: such as Frank Izaguirre of Pittsburgh, Pennsylvania, who gave my book on bird names such a generous review. We have been sharing our warbler sightings this spring – though I do feel that his constant parade of colourful American wood-warblers rather trumps my chiffchaffs and blackcaps. What is lovely, though, is that we are sharing our common interest in spring migration, despite being thousands of miles apart.

One day earlier this month, someone posted a suggestion for nature-related names of James Bond films. As often happens on Twitter, suggestions came thick and fast: You Only Live Twite. Goldfincher. Corncraker. Casino Royal Tern. My personal favourite: The Man with the Golden Plover. And, of course, a genuine film title, based on the name of Ian Fleming's home in Jamaica: Goldeneye.

Twitter can be life-affirming, entertaining or just plain annoying: one morning, I saw that one of the juvenile white-tailed eagles recently reintroduced to the Isle of Wight drifted over Taunton, then up the M5, before turning east and passing over the Somerset Wildlife Trust reserve at Westhay Moor – which means it may well have flown right over our house, or at least close to it. Given that white-tailed eagles are roughly the size of a Zeppelin (well, I exaggerate slightly), I am sure I would have seen it had I been following my friend David Lindo's sage advice to 'Look up!'

Week Five

*   *   *

On the morning of Tuesday 21st April, I awoke at what a friend of mine calls 'stupid o'clock'. Unable to get back to sleep, I decided to do an ad hoc survey of my new patch, while taking a (very) slow bike ride around the loop.

During the three-mile, thirty-minute journey I tallied a very respectable fourteen species of singing or calling birds, made up of sixty-two individuals. Having written a book on Britain's most common bird, I was not unduly surprised that almost forty per cent of the total – twenty-three in all – were wrens, each defending their tiny corner of Somerset with their explosive, frantic song.

The second commonest was the robin, with nine, followed by blackbird and goldfinch (five), and skylark, great tit and chaffinch (four). I also heard my very first sedge warbler of the year, just back from West Africa, which twice unleashed a torrent of song from the depths of a bramble, though to my frustration declined to show itself.

Apart from the rabbits in the garden, which Rosie always finds a challenge, mammals are not a common sight around here – we hardly even see grey squirrels – so I was delighted to come across a fox being harangued by a magpie, and then chasing it back; and a pair of roe deer, which caught sight of me before they bounded off across a newly ploughed field.

One mammal was mercifully conspicuous by its absence: I didn't see a single human being, either walking, cycling or driving, during the whole time I was out.

The next day, 22nd April, I again rose early to carry out my first Breeding Bird Survey of the season for the British

Trust for Ornithology (BTO). This relies on thousands of keen amateur observers – like me – spending an hour or so, twice each spring, counting the birds they see and hear along a pre-determined route; in my case, a two-mile journey through the centre of my home village to the other side.

Again, I usually do this by bike, and this year tallied a typical twenty-plus species, including lots of wrens and robins (again), chiffchaff, blackcap, buzzard, stock dove and a green woodpecker (unseen, but uttering its distinctive laughing call).

I also recorded two familiar species whose common names are similar, yet whose lifestyles could hardly be more different: the (barn) swallow and the (house) sparrow. One is a global wanderer, which has been seen on all seven continents, the other a bird that rarely travels more than a mile from where it hatched out, during the whole of its brief life. This spring, at least, I felt more empathy with the sparrow.

Many people have been unable to carry out their BTO surveys this year because of the lockdown, but I felt it was important to do so: there has been a lot of speculation on social media, and in the newspapers, about whether nesting birds are benefitting from the lockdown (less pollution, noise, disturbance, etc); yet we will only know one way or another if we continue to count them.

*   *   *

Later that morning, again at 'buzzard o'clock', Suzanne noticed a ludicrously distant red kite high to the west, which I could barely see, even through my binoculars. She then excitedly called out, 'Swifts!' Metaphorically donning my 'mansplaining' hat, I prepared to suggest that they were, of course, swallows, until I saw them: three dark shapes scything through the air like guided missiles. These were not only our first swifts of the year, but a full nine days earlier than our previous earliest over the garden, and our only April record in fourteen springs.

No sooner had we identified them and they were gone – continuing east on their epic journey from south of the Sahara to who-knows-where. I tweeted the sighting, quoting the poet Ted Hughes, for whom the swifts' annual return proved that 'the globe's still working...' I did momentarily wonder if, one year, they will not come back – in which case it really will all be over. But I set such grim thoughts aside and spent the rest of the day with a broad smile on my face.

Meanwhile, Suzanne has been enjoying encounters with more familiar birds, too. As she says, seeing blackbirds and robins each day is the avian equivalent of comfort food: like eating a shepherd's pie smothered in tomato ketchup. I *think* I see what she means...

Next morning, the patch produced a veritable buzzard-fest, with a variety of individuals, including dark ones, pale ones, those with reddish tails, and an almost white-fronted bird I nicknamed 'Frosty'. Seeing it reminded me that when Suzanne used to drive the children across the moor to their first school (in the next village but one), Charlie would give names to each buzzard they saw, including the wonderfully evocative 'King Soar'. Now they are teenagers, the children appear to have lost that childhood interest in birds, but I am confident that at some point in later life it will return.

Another, much scarcer raptor appeared this evening across the moor: the second female marsh harrier of the spring (the first was the one Suzanne and I saw on her birthday, the day before lockdown began). Unlike that bird, which flew straight through on a north-east trajectory, this one hung around, giving me splendid views of her chocolate-brown plumage and contrasting pale head and face, bisected with a dark stripe through each eye. She floated effortlessly over the central fields, as if hunting, and I wondered if these birds are not migrants (as I had originally assumed) but have simply wandered over from the nearby Avalon Marshes for a change of scenery.

When we moved here, just fourteen years ago, marsh harriers were very rare birds in Somerset, with only a single breeding pair. Today they are, if not exactly common, virtually guaranteed to appear on any visit to the RSPB reserve at Ham Wall, a few miles to the south.

I'm old enough to remember when there were thought to be just one male and two females of these polygynous birds in the UK, all living on another RSPB reserve, Minsmere in Suffolk. Back in 1973, on a visit to Minsmere, my mother noticed a large, dark bird flying low over the reeds. To my delight, the legendary warden, Bert Axell, identified it as a marsh harrier, uttering the immortal phrase, 'Well done, madam!' Today, thanks to the RSPB and other conservation organisations, there are at least four hundred breeding pairs of these elegant birds of prey in the UK.

I have also reached a lockdown-related milestone. During the whole of last year I managed to tally sixty-two species on this newly acquired patch – which at the time I thought was a pretty good total. Yet this year, I have already surpassed that figure by St George's Day, thanks to the first reed warbler and lesser whitethroat of the year, both singing from deep inside a dense hawthorn bush near the first junction.

Up in Shetland, six hundred and fifty miles away, and nine degrees latitude to the north of here, my friend Donald S. Murray has been writing poetry. I first met Donald more than a decade ago, when I interviewed him for my TV series *Birds Britannia*. From the start I was charmed by his stories, deep Hebridean voice and ready smile, and although we only see each other every two or three years, I regard him as a dear friend. So, I was delighted when he sent through his latest verse, simply entitled 'Quarff, Shetland – 15th April 2020':

At six o'clock (or so) this morning,
they broke out of social isolation,
their songs a mix of celebration,
wooing, warning
as they rejoiced in close proximity
within a garden edged by trees.

Did I envy them – these starlings, sparrows?
Yes – for the way they recklessly embraced
their neighbour's feathers,
squabbled beak to face,
swirled into the distance.

Unlike those of us who've bunkered down,
the grounded human race.

As a simple summary of the gap between the carefree lives of these common and familiar birds, and our own confined existence, this is hard to beat.

*   *   *

On Friday 24th April, we heard some news we had been expecting. Our dear friend Mark – an old colleague of Suzanne's, and godfather to our son Charlie – was to be discharged from hospital that afternoon, after suffering from a truly terrible bout of Covid-19. For the previous ten days or so, after he had been rushed into hospital, we had followed his progress day by day, praying that he would pull through. Thankfully, he had – and now, with only a few hours' notice, he was being sent home.

But as Mark lives on his own we had already decided he would come and convalesce here with us in Somerset. So, that afternoon, Suzanne drove up to west London to pick him up. Late that evening, they returned. It was tough at first, seeing Mark looking so ill and gasping for air as he took the few steps from the car to the house. But once he had sat down and recovered his breath, we could see that he was incredibly relieved to be out of hospital and here with us. It was a timely reminder that this lockdown is not an exercise, but the only way we can stop the spread of this terrible disease.

My significant birthday weekend began the next day, 25th April – St Mark's Day. I tweeted the entry from Wonderland on St Mark's fly (sometimes called the hawthorn fly). In response, naturalist and author Jules Howard informed me that they are called St Mark's flies because the eponymous apostle writes in the New Testament of Jesus smiting the air 'with millions of dancing flies'. I wondered if this is an urban myth, as intense Googling failed to provide any evidence of this quote. I had always assumed the name originated because they appear, with metronomic regularity, on St Mark's Day.

I actually saw my first one about a week ago: an untidy-looking creature, with legs hanging down beneath its body, zigzagging across the garden on a sunny day. Fly expert Erica McAlister won't like me saying this, but basically St Mark's flies are most useful as an abundant source of food for migratory warblers, chats and flycatchers, all just returned from Africa.

* * *

On 26th April, I turned sixty. This was neither as traumatic nor as dull as I might have feared. I have long believed that age is a state of mind, and am constantly surprised at how young most of my contemporaries look – maybe sixty really is the new forty!

Besides, our generation always has the consolation that we are considerably younger than that cohort of musicians we admired as teenagers, and who still behave as if they are about eighteen. This was brought home to me watching Mick Jagger cavorting in his front room on the recent BBC lockdown gig, while Keith Richards strummed his guitar and the uber-cool Charlie Watts played air drums. But I also remember how old my nan seemed when I first knew her, as a grey-haired old lady in her early sixties – things were, however, very different back then.

I had originally planned a party for the night before, with local friends and a sprinkling of those from further afield, which had of course long been cancelled (or, I hope, postponed). Yet by now I had got used to the lockdown and was rather looking forward to the birthday itself, which conveniently fell on a Sunday.

It got off to a great start: I was enjoying an al fresco fried breakfast with David, Kate and Suzanne, when a cuckoo called from somewhere at the bottom of our garden. While I regularly encounter cuckoos on the Avalon Marshes, a few miles down the road, this was only the second time I had ever heard one from the garden – the first as recently as last May. It felt like a good omen.

The cuckoo's call is so close to the human voice, and consequently so easy to imitate that they always sound like a parody of themselves. But they never fail to bring joy to the hearer: I recently got an excited email from a friend in a nearby village, telling me that she, too, had heard her first cuckoo of the spring – a rare event nowadays.

Later that day, I read Rob Penn's lovely article in the Observer, in which he wrote eloquently about hearing the cuckoo, and how this and other signs of spring have spurred him to keep a nature diary for the first time:

> *Observing the changing seasons in a diary is also an internally imposed discipline, a form of spiritual self-monitoring, while the storm rages elsewhere…*

On this special day I had woken early, as I often do now-adays, and decided to announce my birthday on Twitter. I mentioned my mother, who had been a single parent – or 'unmarried mother', as they were called in those benighted days. I also paid tribute to my nan, who had given up her job in a music shop, at the age I am now, to bring me up.

I felt a sense of immense pride, tinged with a wave of sadness, as I tried to imagine the joy and fear they must have felt when I was born, sixty years before. Posting this, I was overwhelmed by the kind comments on Twitter – many from people I do not even know – and was glad I had told their story on this special day.

After breakfast, I went on my daily bike ride – this time a leisurely tour around the moor to take in both the beau-tiful weather and the birds. I love the fact that my birthday always coincides with the peak of spring migration, and today was no exception – paradoxically made all the more special because of the lockdown restrictions. This was my day, in my place, with my birds…

My first stop, after just half a mile, came when I heard a bird that has flown here all the way from somewhere in the heart of north-east Africa – a lesser whitethroat. This is a species which, until I moved down to Somerset, I had only seen or heard a handful of times in Britain. Since then, though, I have learned its subtle but distinctive song, which is just as well, because I hardly ever see them.

As my friend and fellow author Dominic Couzens says, the lesser whitethroat is a prime candidate for the title of Britain's most annoying bird. Not only do they make skulk-ing into an art form, they also move silently and unseen from one part of the thick hedgerow to another, so that just after staring at where I think the bird is for a minute or so, I then hear it right behind me. Seeing it is, if not quite impossible, certainly time-consuming.

Yet despite this, I love the lesser whitethroat. Its song is a soft, almost inaudible warble, followed by a rapid burst of notes, somewhat reminiscent of a chaffinch or a yellowhammer, but more regular in rhythm and tone. If I do manage to catch sight of the singer, it is a slender and very dapper little bird, with a neat, immaculate plumage: grey above, pale below, with a whitish throat and dark mask across each eye.

This particular bird has taken up territory in a patch of hawthorn near the start of the loop and is singing quietly to attract a mate and repel rival males. Normally, I may hear one or two lesser whitethroats around the loop, but this year there are no fewer than six. I hear from other birders, elsewhere in Somerset and across the UK, that they too are hearing more than usual – a small but significant piece of good news.

As I stood and tried to catch a glimpse of this songster, its cousin, the common whitethroat, suddenly launched itself out of the same tree, as if fired from a catapult, to deliver its rapid, scratchy song. These two species may share a name and a superficially similar appearance, but they could hardly be more different in their behaviour: one shy and elusive, the other confident and brash.

You might also assume, given that both species fly here from sub-Saharan Africa, that their winter ranges overlap, too. Yet while the common whitethroat travels from West Africa, like the sedge and reed warblers I could also hear, the lesser whitethroat takes a very different route. It flies here from the north-east of that vast continent, where they overwinter in Sudan and Ethiopia. And once they have finished breeding, lesser whitethroats don't take the usual route south through Iberia, crossing the Mediterranean at Gibraltar, but east, flying through the Middle East (where I have seen them in vast numbers in Israel). Yet another way in which this modest little bird surprises us.

*   *   *

As the days go by, I feel more and more like some modern version of John Clare, the early nineteenth-century poet who spent the first three decades of his life in one village, on the edge of the East Anglian fens. The lockdown has given me – and naturalists all over the country – some small insight into what it must have been like to have lived that way: to know not only every bird, mammal and wild flower, but also more or less every tree, every hedgerow and every field, and have that deep-rooted sense of place that so many of us have lost.

I'm also struck by the distances the birds I hear have travelled: the lesser whitethroat from north-east Africa, the common whitethroat, sedge and reed warblers from West Africa, and the swallows, of course, from southern Africa. Clare, too, recorded their return each spring, at a time when people were still only dimly aware of the phenomenon of global bird migration; and he also marvelled at their epic journeys, and welcomed them back, as we all do this special spring, with joy and delight.

Week Six
27th April–3rd May

Wheatear

At 7.30pm on an unpleasantly soggy day – an unfamiliar experience for us so far during the lockdown – I realised that, having foregone my usual early-morning bike ride, I hadn't actually been outdoors all day. So, I put Rosie into her harness – a process that takes a while, as she is always curiously reluctant to be taken for a walk – and we wandered down the lane towards the moor. As soon as we got outside the gate her mood perked up considerably, and she trotted happily alongside me, only occasionally pausing to sniff a particularly compelling patch of vegetation on the verge.

Seemingly overnight – though it must have been happening gradually during the previous week or so – the lanes are now lined with the creamy heads of cow-parsley, broken from time to time by the electric-yellow of oilseed rape, which has floated over from some distant field, to arrive as an unwelcome intruder onto what I now regard as my patch.

The heavy rain from earlier in the day had lightened to an almost imperceptible drizzle, and as if to make up for the earlier silence, the evening chorus was in full swing. I could hear the fluty tones of blackbird and song thrush, accompanied by the baa-ing of lambs from the field beyond, and an unfamiliar smell: the refreshing scent of newly fallen rain on the ever-burgeoning foliage along the hedgerows.

The evening is always a good time to compare the songs of the song thrush and blackbird, as they continue to sing when the other birds have already settled down for the night. I am one of that small minority of people who prefer the repetitive, insistent phrases of the song thrush to the more mellow, yet to me rather dull, sound of the blackbird.

I just don't get why so many people – including, it must be said, my wife and most of my fellow nature writers – find the blackbird so compelling and evocative. Before the lockdown started, I was giving a talk at my local independent bookshop, where I decided to provoke my audience by telling them this. I was actually hissed at – albeit in a good-humoured way – for stating my preference.

The rain had changed the whole complexion of this familiar walk: swallows, which until then had been few and far between this spring, were flying low in loose, noisy flocks, presumably catching up on feeding after a frustrating few hours when the flying insects would have been hard to find.

It felt good to be doing the walk without the usual little groups of people: last Sunday these lanes were, if not quite like the proverbial Piccadilly Circus, noticeably more populated than usual. But at this late hour, they were completely empty. For someone who grew up in the London suburbs, and who has always lived in cities – at least until we moved here back in 2006 – I am growing surprisingly tolerant of my own company, and that of the dog.

As we approached the final junction, half a mile from home, I could hear, even at distance, the repetitive, grating song of the reed warbler – sounding rather like a less tuneful version of a song thrush. The first time I ever heard a reed warbler here was actually in our garden, the June after we moved to Somerset: a male singing its heart out in the lilac bush outside our home.

It had felt as if nature had come to me in a most unexpected way; only later did I realise that reed warblers breed in the clumps of vegetation in the rhynes around the lanes, and not just in the vast reedbeds down on the levels themselves. I wonder if these are the birds that haven't managed to secure prime territories, and have to make do with this second-best habitat; or maybe the species is more adaptable than we think, and these thick, tightly packed hawthorn hedges above tiny trickles of water are ideal for them to raise a family. I rarely, if ever, actually *see* reed warblers here; only in July or August do the newly fledged youngsters climb to the tops of the tiny patches of reeds and momentarily show themselves.

Rosie seemed to be enjoying the unfamiliar sensation of wet grass beneath her feet and her nose. She padded along beside me, constantly sniffing, occasionally pausing, and always alert to anything unusual along this now familiar walk.

Just as we neared home, a lone common whitethroat shot up into the air and uttered its distinctive loud and scratchy song, which I always think sounds like someone counting rapidly from one to seven, very different from the more regular tones of its lesser cousin. These two sounds will be my constant companions for the next few weeks of spring.

One of my first recollections of reading a serious ornithological paper was back in early 1974, in the BTO's journal Bird Study, with the striking headline 'Where Have All the Whitethroats Gone?' I'm not sure if this was a deliberate echo of the Joan Baez protest ballad Where Have All the Flowers Gone? but it certainly had a profound effect on me.

The authors sought to explain the massive population crash of common whitethroats in the spring of 1969, when more than nine out of ten of these elegant warblers failed to return to Britain from their wintering grounds in Africa. It turned out that the cause was a severe and sudden drought in the Sahel zone, the narrow strip of green and fertile land between the Sahara Desert to the north and the tropical jungles to the south, where whitethroats spend the winter.

I had only seen my first whitethroats in 1973, while birding with my friend Daniel in Bushy Park on the outskirts of London; I still considered them a rare and exotic species, so had not noticed that they had suffered such a catastrophic decline. But this was, I realise now, the first time it had ever occurred to me that bird populations were not somehow stable and immutable; that they could suffer declines following unexpected events such as the Sahel drought.

From the perspective of half a century later, we can now also see that this utterly unexpected event, as this hitherto green and fertile land turned to dust, was one of the first inklings of what would ultimately become the global climate crisis. Had we but known, perhaps this early warning from an overlooked little bird might have allowed us to escape the consequences we face today.

So, although hearing – and momentarily seeing – this particular whitethroat brought a spring to my step as darkness fell on that damp late April evening, I was only too well aware that this little bird's survival still hangs on a very fragile thread.

*   *   *

Earlier that day, we had celebrated the birthdays of my son James, happily marooned in Japan, and my half-sister Elisabetta, less contentedly cooped up with her husband and two young daughters in Lombardy, the epicentre of Europe's coronavirus outbreak. We linked up with James, his mother and stepfather, and his eldest brother Daniel's family, and had a long chat about how he is coping. Fortunately, as a teacher of high-level English, he has been able to move his work online, so is not suffering financially from the crisis. What was more worrying was that the Japanese government's response seems so vague and unfocused that it can only be a matter of time before the virus takes hold there.

The next morning, I awoke early and wondered about

the wisdom of heading out on my bike, especially as a heavy layer of mist hung over the fields, obscuring all but the closest view. But having dressed up in my Lycra (yes, I am that kind of cyclist) I decided to continue, and I was very glad I did.

Not only did I hear and see two splendid sedge warblers, frantically chattering from the banks of the rhynes, but I also heard my second cuckoo of the spring, calling a couple of times way over towards the village of Blackford. One sedge warbler sat out right in front of me, showing off his pale ochre underparts and distinctive yellow flash above his eye, as he opened his bright red gape to sing. It felt as if a little piece of Senegal had come to Somerset.

At the start of the route I also bumped into our next-door-neighbour-but-one, Malcolm, and stopped briefly for a socially distanced chat. As a self-employed builder, Malcolm is currently struggling for work; by contrast, his wife Bev is a nurse at the local hospital, so at the frontline of this crisis. Once again, I was glad of the human contact; something I have come to value so much nowadays. It had truly been well worth coming out on this rather dank and misty morning.

* * *

May Day dawned bright and sunny, continuing the pattern set by April – which as the Met Office later confirmed, was the sunniest on record across the UK, with an average of over seven hours a day. If anything, Somerset enjoyed even more unbroken sunshine.

My early-morning bike ride produced more warblers than any other time this spring: a total of sixteen singing males of five species, plus an added bonus in the form of two fine male wheatears hopping about in a furrowed field just before the swans' nest.

Wheatears – the name comes from the Anglo-Saxon meaning 'white arse', a reference to the bird's conspicuous white rump – are one of Suzanne's favourite birds, so later on we both popped down to see if they were still there. To my delight, they were, showing off their perky stance, dove-grey and ochre-yellow plumage, and black bandit-mask across the eyes. However, by the next morning they had departed on the continuation of their journey north, from West Africa to northern England, Scotland, or perhaps even across the North Sea to Scandinavia.

While some birds, such as the house sparrow and tawny owl, live here all year round; and the whitethroats and other warblers come here to breed; others, like the wheatear (and a flock of whimbrels I saw here ten years ago, on the morning of my fiftieth birthday), simply pass through each spring and autumn on their twice-yearly global voyages. There was, as always, something very comforting about witnessing this, even momentarily, and especially in Suzanne's company.

It had been seven weeks since I last had a haircut, and so later that day, having located a set of clippers on the Internet (almost as difficult as finding a bag of flour in these strange times), I foolishly allowed Daisy to have a go at giving me one. The resulting gales of laughter were only stifled when her brother Charlie took pity on me and insisted on finishing the job himself. The resulting 'buzz-cut' (what we used to call a crewcut) was certainly severe, but thanks to Charlie, at least it didn't make me look as if I had wielded the clippers myself.

*   *   *

Sunday 3rd May was International Dawn Chorus Day, and I was delighted to read a feature on conservationist Mark Avery's blog revealing the origins of this global event. My old friend and mentor Chris Baines explained that, back in the late 1980s, he had chosen to celebrate a significant birthday by organising a communal dawn chorus event near his home on the outskirts of Birmingham. The following year this was picked up by the media, and BBC Radio 4's *Today* programme announced that the first Sunday in May was International Dawn Chorus Day, which it has been ever since.

I have to confess that, having woken at 4.45am to a blackbird (of course) singing just outside our bedroom window, accompanied by the sound of rain beating down on the roof, I decided to stay in bed this year. But all around the world, people did take part, posting their sightings (or should that be 'hearings'?) on social media.

I was especially pleased to hear from my old mate Sean Dooley in a suburb of Melbourne, whose first bird (and, with Australia so far ahead of us, time-wise, one of the first reported in the world) was the charismatic Australian magpie, which, as he tweeted, 'seemingly stirred noisy miner, rainbow lorikeet and grey butcherbird into action'. I do love Aussie bird names.

Skylarks with Rosie

Later in the day, our friends in the States tweeted a veritable avalanche of birds, including such wonderfully named (and equally beautiful) gems as indigo bunting, cerulean and prothonotary warblers (the latter supposedly named after the yellow uniforms worn by some long-forgotten Vatican official).

The whole event was one of the most uplifting global happenings I have ever witnessed, and a real boost to us at such a strange and difficult time. For me, our shared love of birds always has that effect – making the world seem both infinitely bigger and infinitesimally smaller at the same time. As Sean himself wrote:

Birds link sky and earth, they migrate across hemispheres, traverse the vast oceans… In our noisy, industrial society, birdsong has largely been drowned out, all too easy to ignore. However, in the quietude of lockdown and isolation many of us have once again become attuned to their sound breaking the silence.

The difference is, confined to our homes all day, surrounded by a quiet most of us have never experienced, we are noticing for the first time the daily soap operas of the feathered characters we share our surroundings with.

On Sunday I will be listening for these comforting, yet bittersweet sounds – palpable reminders of the natural world I am cut off from, while at the same time providing me an essential connection to the world I once knew, the world to which I want to return.

Being a Sunday, I didn't have a lot to do; so, after reorganising my bookshelves – again – I watched an as yet unreleased documentary made by two young wildlife film-makers, Luke Massey and Katie Stacey. The Last Song of the Nightingale soon got my undivided attention, and not just because I am an on-screen contributor.

This was the compelling story of a bird which, despite our long-held fascination with its extraordinary song, is now in such a serious decline that it may well disappear as a British breeding bird in the next couple of decades. It was both uplifting and terribly sad to witness both the bird itself and the people who care so much about its fate – including the nature writer Richard Mabey, musician Sam Lee, TV presenter Chris Packham and rewilding guru Isabella Tree.

It is such a paradox that we Britons care so much about the natural world, yet we are, as Isabella pointed out, almost at the bottom of the league table of wildlife-rich countries. At a time when, in the run-up to the seventy-fifth VE Day anniversary, we hear so much about how wonderful we are as a nation – what one commentator has called 'the myth of British exceptionalism' – it makes me ashamed to have our failure to save one of the most culturally important wild creatures on the planet revealed so starkly, and yet so tenderly.

I sincerely hope that Luke and Katie's wonderful film gets the audience it deserves. But I am not holding my breath: British broadcasters are always keener to show the antics of penguins or the usual parade of primates and pred-ators than they are to confront us with the truth about the wildlife declines in our own backyard.

Week Seven

4th–10th May

Pipistrelle

On the last day before the rescheduled Early May Bank Holiday, I finally finished work at around 8pm. After a few games of table tennis with the boys (during which I, as always, was soundly beaten), I decided that as the weather was so good, and the forecast for the next morning so bad, I just had to go on an evening bike ride around the moor. As I set out, the sun was just sinking towards the horizon: that clichéd ball of fire in the western sky, towards the landmark of Brent Knoll, one of the few high points in this relentlessly flat landscape.

A day or two earlier, I had been talking to Chris Packham, and he asked if I had noticed that the volume, intensity and variety of birdsong had started to decline, especially amongst resident species, which start to breed earlier than their migrant counterparts. I hadn't thought about it till then, but he was quite right: thanks to the long spell of very fine weather, most of our commoner breeding species have now got chicks in the nest, and, as the males join in the feeding duties, they have less time and energy to sing.

But once dusk starts to fall, they seize the chance to lay claim to their territories once again. On my leisurely pedal around the loop I could hear wrens, robins, blackbirds and song thrushes, all singing their hearts out.

It isn't just the lengthening evenings that tell me that spring is infinitesimally shifting into summer: the foliage of the hedgerows and verges is almost bursting its banks, with lanes lined with cow parsley, the unwelcome splash of yellow rape, and even the occasional blooms of red valerian along the walls in the village – a plant I usually associate with June.

These, and the gradual decline in birdsong, are signs that that intense and oh-so-exciting burst of energy during the last weeks of April and first week of May is now almost over, giving way to a subtler series of changes. Nature is beginning to settle into a comfortable groove, as if spring has embraced middle age.

And, indeed, this week has marked the halfway point of the thirteen-week season of spring; a time that always seems to go too fast, and never lasts quite long enough. At least this year I have felt I can appreciate it, watching it unfold day by day and week by week, in a way I cannot recall doing for years – perhaps ever.

We have shared this experience with friends all around the northern hemisphere, via the lifesaving social media channels. Our friends in South Africa and Australia are reporting longer nights, in that strange juxtaposition of opposites, in which half the globe is welcoming summer, while the other half experiences the onset of winter. Again, this is something we have all known about all our lives; I just cannot recall ever feeling it so clearly, or so intensely, before.

I certainly don't think that, in the fourteen years we have lived here, I have ever truly appreciated the variety of birdlife on the moor. I have perhaps ignored it because it is so close to my home; and because it does not have the areas of open water of the nature reserves to the south, which attract a wider range of species and so provide more excitement for the visiting birder. Yet there is something to be said for ultra-local birding. As I listened to a lesser whitethroat chuntering unseen in the hawthorn hedge, against a backdrop of so many other species, I realised that I shall never be able to look at this place in quite the same way again.

Home is a strange concept. Ten years ago, if you had asked me where my home was, I might have struggled to answer. I was brought up for the first eighteen years of my life in the same small semi-detached house in the London suburbs; then after three years at university in Cambridge, I lived in various parts of west, north and south London.

Coming down to Somerset in the summer of 2006 was a huge step. We knew that this would be our final move, and it has proved to be just that: we are settled in our home, in our village, and in this beautiful and underrated county, and will remain here in this house for the rest of our lives.

This spring, though, our enforced confinement has made us realise just how lucky we are to have settled in such a place: not the most beautiful, the most scenic or the most dramatic of landscapes, but most definitely home. The thought that in a decade and beyond, I shall still be tramping these lanes with a much older Rosie fills me with a deep sense of comfort and joy. Another cliché, of course; but then sometimes clichés are just what a writer needs.

As I turned the corner for home, the sun had finally set, and the sky was showing off that panoply of colours that defy description. I thought of my friend Brent, in lockdown in South Africa; we only got to know each other a few months ago, when I went to see the swallow roost. Brent is an artist, with equal measures of talent and modesty, and as soon as we met I felt that we had been friends for years.

One evening, as we drank sundowners at the end of a safari drive, he challenged a fellow guest to a painting challenge, to capture the landscape as the sun rapidly set behind a nearby hill. As he painted, with superb dexterity and speed, he talked: telling me about his theory that painting is first and foremost about *seeing*, then *looking* and finally, above all, *feeling*. 'I tell my students to always catch the moment, and to not just look, but feel. Purple is the key – it exists everywhere, in terms of the shadows; all the shadows are a different type of purple, you just need to find them.'

A few weeks after I had returned home, a package arrived. Opening it, I was touched to see the painting Brent had been doing that same evening, which I had watched him create from nothing. It now takes pride of place in our home, a reminder of the need to see the world in new ways, whether you are a painter, poet or, like me, just a nature writer.

Perhaps 'feel the purple' is a metaphor for life at the moment. Everything we see, hear or feel has always been there, but for the very first time we are genuinely noticing and appreciating it. A simpler, clearer way of living. There has been a lot of talk in the mainstream and social media about how we mustn't go back to how things were; that this terrible crisis has given us a genuine, once-in-a-lifetime opportunity to reset the world and the way we all live our lives.

It's a tricky dilemma: not everything about the old world was bad by any means. But what was wrong was our long-standing inability to understand that we are not separate from nature, but inextricably part of it. If we can just change that, then this horrible time might at least have some positive consequences.

*   *   *

After I returned home, I noticed a movement outside our sitting room. Half a dozen bats were shooting around our back garden in the gathering gloom, just above head height. I managed to recall where I had put the bat detector I bought at Birdfair a few years ago; fortunately, the batteries were still working.

Turning it on, I was immediately assailed by an extraordinary sound, retuned, by this little machine, to a frequency the human ear can hear. The bats were echolocating: emitting a series of clicks that bounce off any nearby surface, allowing the creatures both to navigate in the darkness (and avoid bumping into any obstacles) and to hunt down the unseen moths flying around and above the garden.

Soon after we moved here, I went up to the top floor of the 'workshop' – a building then attached to the side of our home. On the floor, amongst the usual junk and detritus, I found a pile of large yellow underwing moths – minus their bodies – the victims of a particularly efficient bat.

Many years later, when we had the workshop demolished to build our new extension, we spent a ridiculous amount of money carrying out bat surveys, to ensure that none would come to any harm. A council officer turned up to observe, and I asked him what would happen if we found any bats. He replied that he would simply halt the work for a few minutes, capture the bats and put them in a box to take elsewhere and release. We did not find a single one. The whole palaver did make me question why, in that case, we had to have such comprehensive surveys done. The only benefit, as far as I could see, was that we discovered that at least seven species of bat used our garden – but at around £600 per species, that seemed quite a high price to pay.

On that fine evening in early May, the bats were soprano pipistrelles, which have quite a story to tell. Pipistrelles are our commonest bat species and were always known to echolocate at two distinct frequencies: forty-five and fifty-five kilohertz. Then, in the 1990s, scientists discovered that this was not a single species, calling at two different frequencies, but *two* completely different species, echolocating at one each. The two new species were named common pipistrelle and soprano pipistrelle – the latter calling at the higher pitch.

The next night, the whole family came out onto the

veranda to see and hear the bats. It was a lovely moment, following which we noticed a bright shape in the western sky. Getting the telescope out, we gazed at the planet Venus, showing a strange crescent 'new moon' shape, which is apparently the result of its position relative to the Earth and sun. I realised that we would never normally take the time to look at either the bats, a few feet above us, or Venus, almost thirty million miles away. Yet, because of the way lockdown has rekindled our family bonds, we are all enjoying both these spectacles together.

*   *   *

One of my favourite places around where I live is Tealham and Tadham Moors – two contiguous pieces of land south of the Isle of Wedmore that flood in winter and so are packed with bird, flower and insect life each spring and summer. Though they are farmed sympathetically for wildlife, they are not actually nature reserves, which makes them all the more precious. In some ways, this enhances the experience of visiting them: there is a real sense of what much of lowland Britain could be like, if we just made more room for nature.

I say 'visit', but the usual way I see the moors is when driving across them, as they are a useful short cut to avoid going through Wedmore on the way to Glastonbury and the Avalon Marshes. But that doesn't preclude seeing wildlife: last winter, as I drove back from a shopping trip with my binoculars in the car boot, I noticed two small raptors duelling with one another just above the grassy area to my right. Leaping out and grabbing the binoculars, I was able to confirm that though the pursuant was a kestrel, the smaller bird was a merlin, which fortunately then sat on a fencepost and allowed me to watch it for a full twenty minutes, before I had to head back home.

I've seen other memorable birds here, too: it was the first place I ever saw cattle egrets in Somerset, and I can still recall how excited I was, as I hadn't heard the news that they were around. Other highlights over the years have included wood sandpipers, whooper swans, short-eared owls and flocks of whimbrel passing through in spring.

So, when I heard that not only was the fine weather going to continue into the weekend, but also that a local café at the eastern edge of the moors was open for take-away bacon rolls and coffee, I was keen to take a bike ride over there; our first proper outing since lockdown began. David and Kate immediately agreed to come, as, with a little persuasion, did Daisy. So, just after lunchtime on a fine, sunny VE-Day Bank Holiday, delayed by only an hour or two by Daisy's intricate dressing and make-up preparations, we set off.

Cycling is such a pleasure during lockdown. Not only are there hardly any cars, but those that we do encounter slow down and pass us in what can only be described as an apologetic manner, as if they are afraid we will call them out for driving at all.

Given that my three companions do not fully share my love of birds and wildlife, I was reluctant to stop at any point on the eleven-mile round trip; though I must confess that the sight of a splendid male yellow wagtail perched on a gate did make me hesitate momentarily. My friend Graeme once described these intensely golden birds as 'like a fucking canary' – perhaps the perfect one-line description of this beautiful but increasingly scarce summer visitor.

On our way back, I heard both Cetti's and garden warblers, meaning this was a 'nine warbler day', if I included my early-morning excursion around the loop. I also saw a damselfly, my first banded demoiselle of the spring, newly emerged as an adult from its underwater lair; and the first red admiral, which had travelled rather further, having flown all the way from Spain or North Africa.

The wildlife was, however, a bonus. Getting the chance to cycle somewhere different, with three of the people I love the most, reminded me not just what the world was like before the lockdown, but also what life was like when I was growing up, a time when you could happily cycle along country lanes without fear of meeting some lunatic driver coming in the opposite direction. Wouldn't it be lovely if that was the case not just now, but in times to come?

* * *

Fuelled by the joys of the previous day, the next morning I decided to visit my usual local patch: not Blackford Moor, which has served me so well this spring, but a hidden corner of the Avalon Marshes a few miles south of my home.

I usually come here with Graeme, on a Saturday morning when our families are not yet up and requiring our services as breakfast-makers, washer-uppers and chauffeurs. Of course, we have not been able to do so since lockdown began, but we still regularly speak on the phone, comparing notes on what we are seeing and hearing. Graeme had texted me the night before to brag about seeing no fewer than ten species of warbler before 9am; so, with that classic competitive spirit that lurks in the soul of all serious birders, I had to try to beat him.

One small problem: as I headed through my parish towards my destination, mist rose up from the flat lands, after a cool night. I wondered if the birds would be singing, and if they were, would I be able to see them? I decided to press on, given that, in any case, I usually identify warblers by sound.

By the time I got to the patch, the mist had begun to lift, to reveal what promised to be yet another beautiful sunny day. The peat diggings by the reserve were home to a lone great crested grebe, a dozen mute swans – presumably non-breeding younger birds – and a pair of Canada geese with six fluffy yellow goslings.

As I walked round, a bulky brown bird flew up from the adjacent reedbed, giving me my first view this year of a bittern – or as one young lad memorably called it, a 'toasted heron'. Bitterns went extinct in Britain in the early twentieth century, and almost did so again in the 1990s, when there were barely a dozen males in the whole of the UK. They really are the ultimate conservation success story around these parts: from a single male in 2006 – the year we moved here – there are now over fifty. And while we don't usually see them, we do hear their distinctive booming call throughout the spring and early summer.

A good start, and it just got better and better. A cuckoo

called and eventually showed itself, first in flight and then perched on a distant bare tree; again, the first I have actually *seen* this spring. But my target was warblers, and over the next three quarters of an hour or so I gradually ticked them off, one by one, by their song: the chiffchaff singing its name, the blackcap sounding like a speeded-up blackbird, the silvery descant of the willow warbler, the chuntering reed and excitable sedge, and the unmistakably loud outburst of a Cetti's.

Six down, four to go; soon to become two, with a common whitethroat launching itself from the little patch of brambles where he – or his ancestors – nest every year, followed by the rapid song of a garden warbler, sounding rather like a skylark crossed with a blackcap.

The ninth warbler species is a rare visitor here, or anywhere in Somerset. The grasshopper warbler sounds, as its name suggests, very like an insect, or perhaps a fishing-reel being let out at speed. And, sure enough, as I approached an area of scrub growing in the middle of the main reedbed, there it was: a few bursts of that unmistakable metallic song.

Strangely, I have never seen, or even heard, lesser whitethroat on this particular patch, and have come across it only once in the whole of the Avalon Marshes. So, to bag my tenth warbler of the day, I headed back to the loop, where one was obligingly singing its heart out in the usual spot. Ten warbler species by 8am. I sent Graeme a gloating text. Sad, I know, but very satisfying. After all, he would have done the same to me.

Week Eight

11th–17th May

Magpie

On the Sunday night before week eight of the lockdown, the Prime Minister spoke to the nation. For once, that was no exaggeration: with almost twenty-eight million viewers, the broadcast shot straight into the top ten of most-watched TV programmes ever, beating both the 2012 Olympics opening ceremony and the Wills and Kate wedding, though falling short of Princess Diana's funeral and the record-holder, the 1966 World Cup Final.

Sadly, the content of the (pre-recorded) broadcast was not quite as historic as any of those events: the PM looked nervous, shifty and tired, and his messages were unclear and confusing, as the almost entirely negative press response the next day confirmed.

At times like this, nature may prove to be our only comfort and solace; though because I have been quite busy with work this week, I neglected my daily trip around the loop until the Wednesday. That morning, as I arrived at the second junction, I noticed a woman peering into the rhyne, which at this point is hidden behind a low, thick hedgerow. As I suspected, she was looking for the cygnets, which had finally hatched out from the swans' nest I had been patiently observing for the previous six weeks.

Moments later, the wildlife cameraman Mark Payne-Gill, who lives in the next village, also arrived. He is filming the swan family – or at least he would have been, if they were here. Fortunately, later that day the parents did return with four beautifully fluffy cygnets and gathered on and around the nest, allowing me to take some delightful photos, and Mark to get some footage for the CBeebies TV channel. This time of year does appear to be hatching time: elsewhere, on the east of the moor, a female mallard was carefully shepherding no fewer than eleven tiny ducklings across the duckweed-covered rhyne.

*   *   *

One indication that spring is gradually shifting into summer is the timing of the first cutting of the grass: not on my own lawn, but in the working landscape of the fields on the moor. One evening this week I took another after-supper bike ride, and as soon as I turned the corner at the bottom of our lane, I was confronted with what looked like giant droppings deposited by some space-alien rabbit. They were, in fact, neat, smooth plastic bags full of silage – each roughly three metres long and two metres in diameter.

I could also smell the newly mown grass, which over the next few days attracted large flocks of birds: including several hundred rooks, whose sharp, pointed bill is ideal for poking into the exposed ground to search for worms, beetles and the other invertebrates (and their larvae) exposed by the cutting of the grass.

They were joined by a dozen buzzards, also searching for food, and a scattering of herring and lesser black-backed gulls, all accompanied by a more than usually tuneful reed bunting, repeating its staccato song. To me, this always sounds rather like a bored sound-engineer calling out, 'One. Two. Testing,' or, if you prefer, 'Eat. My Liquorice.' While I was watching, another pair of mute swans flew low overhead, their wingbeats making that strange, whooshing sound which the writer Paul Theroux famously compared to a couple making love in a hammock.

This week, the lockdown rules have been marginally relaxed, and I headed a few miles down to the coast to meet my birding companion Graeme. It felt good to be resuming our habitual rendezvous, albeit in separate vehicles and a safe distance apart. We had chosen to visit our coastal patch: the confluence of three rivers – the Huntspill, Parrett and Brue – which merge into one and then emerge into Bridgwater Bay.

There has been much debate over the last week about the wisdom of the government's rather confusing advice. People are baffled not just by the widely derided 'Stay Alert' message, but also the fact that, in England at least, we are theoretically now permitted to drive several hours away from our home to enjoy our now unlimited daily exercise.

As I listened to a radio news bulletin, in which tourist officers were pleading with people not to congregate in large numbers at well-known beauty spots, I reflected that at least Graeme and I did not need to worry about that: no one in their right mind would consider a view of Hinkley Point nuclear power station, Steart Point and the faded resort of Burnham-on-Sea to be a beauty spot.

While I waited for Graeme to arrive, by the bridge across the River Huntspill, a woman jogging by paused to tell me that she often sees a kingfisher here. I reflected, as she headed off, that a couple of weeks earlier she and I might have exchanged guarded greetings, but perhaps not actually stopped to talk. Another sign, perhaps, that we are relaxing a little more than we did at the start of lockdown.

Warblers were once again the main attraction, as they are everywhere this spring: seven species, each singing their distinctive songs, though rarely showing themselves. The tide was out as far as it ever goes here, revealing acres of brown mud, on which scattered groups of oystercatchers, shelducks and curlew were all feeding.

We thought we might be a little late for the curlew's smaller relative, the whimbrel, as most of these charismatic waders pass through here in late April; yet to our delight there were at least a dozen feeding on the foreshore, including one that allowed us to get excellent views of its distinctive decurved bill and stripy head-pattern.

I have a real soft spot for whimbrels: having seen them on their breeding grounds in Iceland – where these were headed now – and in their winter home in The Gambia, I always feel as if it is a privilege to watch them as they stop off to refuel here on the Somerset coast.

Another bird it is always a pleasure to see also appeared, as we walked back to our cars: a male sedge warbler, sitting right out in full view as it clung to a reed stem, singing its scratchy, whistling and excitable song in the milky, early morning sunshine. Few other migrants are quite so attractive as a sedge warbler: its pale, creamy-yellow underparts offset by a streaky back and prominent eyestripe, and that wonderful song. As if performing an encore, he then launched himself into the air for his final coda, before plummeting down into the reeds, and out of view.

It was good to be here again – not just to see the birds, but also to catch up with my old friend.

* * *

Later that weekend, there was a very civilised debate on Twitter about what the new government guidance on travelling actually meant, especially with regards to the pastime of twitching – travelling, sometimes for very long distances, to see a specific rare bird. Some, including my friend Lucy McRobert, argued for a more relaxed interpretation of the rules; others suggested that we should continue to be very cautious and still stick to our local patch.

I could see the arguments from both sides: I am lucky to have my 'very local' patch (my garden and the loop) on my doorstep; and two others within a ten-minute drive, or twenty-minute cycle ride. But I appreciate that for many birders, the attractions of a visit to more distant birding hotspots are very tempting, especially when a rare bird shows up.

Interesting, too, that this debate is also being framed in terms of the global climate crisis, and the growing need to use our cars less and less. Maybe twitchers with an electric vehicle should be allowed more leeway than those of us still with petrol or diesel ones? Just a thought…

Meanwhile, we were – as they say – making our own entertainment. Like characters in an Enid Blyton novel, we organised two major family events for the weekend: a Saturday night quiz (complete with takeaway curry – another lockdown first) and a Sunday afternoon Sports Day. Both were the brainchild of David and Kate, whose presence here has transformed our lockdown, making it a really special family time.

Having said that, the Mosses are known for being highly competitive and even more talkative, so both events had their tense moments. Nevertheless, they were also a huge success: yet another way in which lockdown is forcing us to rediscover pastimes that, in our normally busy lives, we would never have found the time for.

I fervently hope that, once this is all over, we do not forget what fun we had answering questions on geography from George, Disney films from Daisy, and complex GCSE algebra from Charlie; and also by competing in musical chairs, obstacle races, and 'eat-a-doughnut-on-a-string' games. We are also enjoying having our friend Mark staying with us, as he convalesces from Covid-19: he was still not well enough to take part in the Sports Day, but he did act as a referee, brooking no argument and keeping us firmly in our place.

Week Eight

*　　*　　*

Bill Oddie always used to say that he finds black-and-white birds as visually interesting as more colourful ones, and I know exactly what he means. There's something about the clean lines of a monochrome plumage that – just like black-and-white photography – can be really aesthetically pleasing. And that is even more so if the bird also shows a splash or two of colour, if the 'black' is actually a subtler shade of very dark blue, or if it has an iridescent sheen when catching the light.

So, I was pleased – if a little puzzled – when, a few weeks ago, a pair of magpies started coming to the tubular bird feeder outside our kitchen window. At first, they looked rather clumsy as they tried, often unsuccessfully, to hang on to the tiny perches designed for much smaller birds. But, after a while, they started to get the hang of it, and soon were happily munching on the seed mix it contained. In the morning sun, from just a few feet away, I was able to really appreciate the subtle beauty of their plumage: the 'black' on their wings being transformed into a mixture of dark green and vivid purple, depending on which way they turned.

At first, I thought nothing of this behaviour, even though I couldn't immediately recall seeing magpies on our seed feeders before. Then, nature writer Tiffany Francis-Baker – who is self-isolating as she is in the third term of her pregnancy – posted the following on Twitter:

Was delighted how quickly the fat balls were disappearing, thinking it was feeding all the little birds in our garden. Turns out it's a massive chubby magpie.

Reading this, the penny finally dropped. The magpies are only resorting to bird feeders because their usual food – roadkill – is in short supply. Responding to Tiffany on Twitter, I was gratified when others confirmed that they, too, had noticed the strange phenomenon of magpies on bird feeders.

And yes, I do know that magpies kill and eat baby birds, seizing them from garden nests, often in full view of the helpless parents and to the even greater distress of the householders watching the carnage from their window.

But I would say, in the magpie's defence, that another black-and-white bird wreaks just as much havoc on song-bird nests, even pillaging nestboxes. Yet no one ever calls for a cull of the great spotted woodpecker – presumably because it does all the damage out of sight, without the noise and fuss made by magpies. And, after all, magpies have chicks to feed, too; indeed, a few days later, Suzanne pointed out a well-grown baby magpie on our back lawn, its parents in close and careful attendance.

We have great spotted woodpeckers nesting close by, too; probably in the nearby orchard where their larger cousin, the green woodpecker, bred last year. Both these birds are usually heard before they are seen: the green giving its distinctive laughing call, and the great spotted uttering a metallic 'pic' sound as it flies overhead.

Whenever I sit outdoors, every half hour or so a great spotted will appear, bouncing in flight from one side of the garden to the other; then returning a few minutes later, presumably with food for its chicks. Sometimes, it comes close enough for me to see the red patch beneath the base of the tail, or the red on the head that confirms it as a male. I recall my excitement at seeing a single great spotted woodpecker in our previous garden in the west London suburbs, and once again remember that I should never take the presence of these beautiful birds for granted.

The third monochrome species flying over the garden from time to time is the house martin. These are never as common a sight as the swallow here; partly because they nest no closer than half a mile away, in the centre of the village, and partly because the species is less common than its more elegant cousin, and suffering a more rapid and precipitous decline. When they do appear, I love watching them as they twist and turn in pursuit of invisible insects in the skies above.

Three 'black-and-white' birds – the magpie, great spotted woodpecker and house martin – all fascinating, and visually appealing, in equal measure.

Week Nine
18th–24th May

Skylark

There's something very different about doing the loop not just at different times of day, but at a different speed. This morning I donned my cycle gear and – in my constant battle to keep fit – went as fast as I could. I did take my binoculars, as an insurance policy after not being able to identify a dark falcon a couple of weeks ago, which may or may not have been a hobby, but this time I didn't need to use them.

But when I go around the loop with Rosie, as I did tonight, I travel at a much slower pace, taking roughly an hour to walk the three miles, as opposed to the ten minutes or so when doing my morning bike ride. The evening hour helped slow me down and unwind from the stresses of the day. It also gave me a chance to take in the commoner birds:

the moorhen moving jerkily across the duckweed in the rhyne at the end of our garden, before performing its vanishing act in the bankside vegetation; a quartet of magpies bouncing energetically around the sheep field; and a pair of goldfinches, which posed momentarily on top of the hedgerow before flying off in a burst of tinkling notes, like those exploding 'Happy Birthday' symbols on emails.

Until we acquired Rosie, I would hardly ever have gone for a *walk* around here: three miles is just too long for an impatient soul like me. I far prefer travelling by bike – you can cover the ground so much quicker than on foot. Yet during lockdown, as with so many aspects of life, I am rediscovering the joys of taking things more slowly, and as a result I am seeing so much more than when I am on my bike. I'm also enjoying keeping this nature diary: not since my first full year here, when I wrote *A Sky Full of Starlings*, have I done so.

Another thing about doing this route so regularly is that I am starting to notice not just each species, but each individual bird. This evening I heard, once again, the distant song thrush that pours his heart out to the north of the moor, accompanied, once again, by the same barking pheasant, also out of sight in the distant trees.

The fields on either side of the lane were awash with yellow: tall-stemmed meadow buttercups, each topped with a cluster of custard-coloured flowers, creating a very pleasing effect as I gazed into the distance towards the setting sun. On either side I could also hear skylarks in stereo: still singing so persistently, despite the gradual shift from spring towards summer.

Skylarks with Rosie

These birds really do seem to sing constantly, for what feels like hours on end, but is probably only a few minutes – another way in which time is slowing down and allowing us to focus on what matters in life. Are skylarks more common than they used to be when we first moved here, or am I just noticing them more nowadays? Given that well over two million pairs of skylarks have simply vanished from our countryside in the six decades since I was born, I fervently hope that they really are beginning to make a comeback.

One reason they are here is because this land is not farmed anything like as intensively as across much of the rest of lowland Britain. Elsewhere, especially in parts of eastern England, the land is worked so hard that there is little or no room for birds like the skylark. There, farming has become industrialised, and is causing as much damage to the landscape as opencast coalmining or fracking – perhaps even more. Yet because we take such a rose-tinted view of the 'countryside', we have become conditioned to see farmers as its 'custodians', protecting it for future generations.

Some, of course, are; but the rich landowners and huge conglomerates that own so much of our agricultural land, whose only motivators are greed and wealth, know little and care even less about the true meaning of the countryside, even as they pocket millions of pounds in taxpayers' subsidies to boost their profits. Skylarks can either be seen as worthless or priceless – and I know which side of that argument I come down on. So, I commend the farmers around here, who manage to eke out a modest living from the land, yet are still somehow able to make room for wildlife.

Week Nine

When I started writing this diary, I wondered if the weather gods were mocking us. But after almost two whole months of dry, sunny and fine weather, I've concluded that they have been on our side all along. How miserable this period would have been if we'd had to endure cloud and rain every day, rather than more-or-less endless sunshine.

As Rosie and I reached the final leg of the loop, I was reflecting on having once again not seen a single soul during the previous hour, when I heard a cheery call from just behind me. It was Steve, my children's former school-teacher, returning home after a long bike ride. We chatted for a few minutes, about life, the virus and everything; and on parting, we agreed to meet for a pint at the White Horse Inn in the village 'when all this is over'.

That's a phrase I find myself using – and hearing – a lot nowadays. Given the continuance of the doom and gloom on the news, I'm never sure if imagining the future makes me feel despair or, as it did tonight, hope.

*   *   *

The lockdown is, it seems, gradually easing. On Friday we heard that George and Daisy, both in Year Ten (the first year of GSCEs), will be going back to school in the last week of June – just over four weeks' time. This won't be anything like normal: they'll be confined to their tutor groups, wear mufti rather than school uniform, and – at least at first – only return part-time. They both received the news pretty well: I think the novelty of being at home is starting to wear off, and they are keen to see their friends again.

For the past week or so they have at least been able to see one friend at a time, provided they stay apart. Charlie took the first opportunity he could to visit his girlfriend Charlotte, near her home in Weston-super-Mare, while George also saw his girlfriend Liv, taking a walk along the Mendips above her Cheddar home. Daisy stayed in the village, meeting a succession of friends, one by one, at the local park. All seem happier for that renewal of precious human contact with their peers.

On that Friday evening we were enjoying a fish-and-chips takeaway around the kitchen table when I realised that on that very day, had the virus not struck, George, Daisy and I would have been arriving in Japan to visit my son – and their brother – James. This 'parallel universe' experience is becoming increasingly frequent the longer the crisis goes on: later in June I would have been on a birding trip to Turkey, while my calendar bears witness to countless talks, birding trips and other appointments, all now cancelled, postponed or being held online. Given how we are still filling our days, I frequently ponder on the elasticity of time: I should have more, yet I still seem to be as busy as ever. That's not quite true: I am finding that my weekends really are a period of rest and relaxation.

This week I continued to broaden my horizons: first with a lovely late afternoon walk around a hidden corner of the Avalon Marshes with Suzanne. Being so hot, it was fairly quiet, though we did see a pair of mute swans with seven cygnets, and at least two male marsh harriers – a bird that Suzanne is currently painting. We also came across a range of 'dragons and damsels', including dozens of matchstick-like azure and common blue damselflies, and a pair of mating broad-bodied chasers, their stout bodies and gossamer wings glinting in the bright sunshine.

But the real highlight came when a slender, long-tailed bird with drooping wings fluttered low across the main reedbed. A cuckoo, yes, but not as we know it: its rich chestnut-brown colour marked it out as a hepatic, or rufous morph female, an unusual plumage variety I had only seen on a handful of previous occasions, and never before in Somerset. A week or so later, I had much better views of this beautiful and distinctive bird, as a male chased it across the reedbed, and on landing uttered its haunting, bubbling call.

The following Saturday, at the start of yet another Bank Holiday weekend, I paid a second visit to the 'three rivers' patch on the coast, again with Graeme. He arrived a few minutes late, so missed the highlight of the visit: a kingfisher, which shot out of the reeds and then flew low along the surface of the Huntspill River towards the sluice. But we did see the last cohort of migrating whimbrels on their way north to breed, and we enjoyed our usual Saturday morning catch-up.

We only met one other visitor: a stout dog-walker with an even stouter Staffordshire bull terrier, who quizzed us about the lack of song thrushes in nearby Burnham-on-Sea. In true Somerset fashion, he came up with a memorable phrase to describe the absence of the song thrush's larger cousin, the mistle thrush, which he told us were now 'as rare as rocking-horse shit'. He's quite right.

That morning, I also made my now regular fortnightly

appearance on the *Today* programme, who like the rest of the media have become obsessed with our new-found national passion for nature. Justin Webb and I discussed the rise of birds of prey: when I was a child, we only ever saw one species – the kestrel – whereas since lockdown began I have seen half-a-dozen different raptors over my garden; the only omission, ironically, being the kestrel, whose fortunes have plummeted while those of the others have risen over the years.

Yesterday, I glanced out of my office window to see a bird shooting low overhead: it was a hobby – my first of the year anywhere. This slender, streamlined falcon looks like a swift on steroids: a Ferrari to the kestrel's Ford Focus. I yelled for Suzanne to come outside, as she loves these elegant predators, but she missed the bird by seconds, as it sped rapidly away to the south.

I spent the rest of Saturday and Sunday in what for me is the unusual occupation of doing some DIY around the house. I say 'DIY', but what this actually entailed was struggling with an adjustable spanner to remove two toilet seats that had rusted themselves well-and-truly in, while listening to the radio station Mellow Magic, whose output is so somnolescent I thought it might have a calming effect on me. It did, and I finally managed to install the two new seats. Afterwards, I felt a strange sense of satisfaction – as I always do when I complete even the most mundane of household tasks.

By Sunday lunchtime, the wind had dropped, and the sun was warming up, so Daisy and I took advantage of the continuing fine weather to do another bike ride. We

crossed Tealham and Tadham Moors with a welcome following wind, and we enjoyed our now regular bacon roll at
Sweet's café. This time we were meeting – at a safe distance
– Richard, the vicar of Wedmore, and his daughter Melissa,
one of Daisy's closest friends, who were also enjoying a
spot of father-and-daughter time. Once again, these simple
pleasures – catching up and chatting about this and that
– seemed so much more enjoyable than they used to be,
when we just took them for granted.

Afterwards, we took the long way back past my local
patch, where a brief stop produced swifts, house martins
and a cuckoo, before heading home through the lanes and
villages of the levels. Even a puncture in Daisy's back tyre
didn't dampen our spirits and, again, my delight in actually
fixing it was wholly out of proportion to the ease of the
task.

What had been a delightful day was then almost spoilt by
the 'Cummings Affair'. This is not the place to go into the
details of this, suffice to say that I had to self-isolate from the
rest of my family and watch the PM's broadcast in another
room, just to avoid shocking them with my frequent explo-
sions of incandescent rage.

Taking to Twitter to vent my frustration at this toxic
combination of arrogance, hypocrisy and incompetence,
I was amazed to see the outpouring of righteous anger
produced by the normally calm, restrained people that I
follow – and who follow me – because of our shared love
of the natural world. I hoped and prayed that this really
was Johnson's 'Poll Tax Moment', when the British people
finally came to realise what a bunch of charlatans are in

charge, and more importantly how badly they have handled this ongoing crisis, lurching from one misguided policy or confused pronouncement to the next, with not a hint of any overarching strategy.

I hoped that would be the case but did not really believe it. Am I becoming (more) cynical in my old age?

Week Ten

25th–31st May

Cockchafer

'Not *another* Bank Holiday!' These were Charlie's words as he emerged at lunchtime on Monday, echoing Brenda of Bristol's famous response to news of yet another general election. I pointed out to him that I wasn't sure why that mattered, given that he was not going to school, nor indeed doing any discernible work. To be fair, he has offered to exercise his design and carpentry skills and make me a desktop bookshelf, though he hasn't actually started work on it yet.

What clearly has started is the Great British Summer: roughly four weeks early, by my calendar. In the words of Phil Gates, in the *Guardian*'s Country Diary column, 'Spring, so eagerly anticipated, so riotous, has sashayed into summer.' Though, as he also points out, 'The natural transition between seasons defies precise measurement. It is the accumulation of countless small events…'

So, while the birds are still singing, almost all are feeding well-grown chicks in the nest as well. This could prove problematic for the BBC's *Springwatch* programme, which, when it began in 2005, coincided with the peak of the bird-nesting season; whereas with the calendar shifting two or even three weeks earlier than the long-term average, many chicks have now already fledged and left the nest by the start of June.

Butterflies are also very few and far between: what lepidopterists call the 'June gap' – that period of time between the disappearance of the spring butterflies such as orange-tip and the emergence of the summer ones like meadow brown – also seems to have shifted forward into May. I did, however, see my very first small tortoiseshell of the year as David, Kate, Daisy and I walked Rosie around the loop on the Bank Holiday Monday afternoon. The next day, one turned up in the garden, flitting low over the unmown lawn and sipping nectar from custard-yellow buttercups in the hot afternoon sunshine.

Small tortoiseshells were, when I was a child, by far the commonest butterfly in our suburban garden, feasting on nectar from my mother's carefully tended flowerbeds. So, like most people, I have always tended to take them for granted. But in the past few years, this handsome butterfly has fallen victim to a parasitic fly, which has extended its range northwards as a result of the global climate crisis.

The fly lays its eggs on the small tortoiseshell's foodplant, which are then eaten by the butterfly's caterpillars. When the fly's larvae emerge, they eat the unfortunate caterpillar from the inside, keeping it alive just long enough for them to grow to maturity. As a result, small tortoiseshell numbers have plummeted in the past few years: yet another example of a creature we once virtually ignored becoming so scarce that it has taken me almost the whole of the spring to actually see one.

Other insects are turning up, though. A few weeks ago, there was a mass emergence of cockchafers – that handsome bronze-coloured beetle also known as the May-bug or doodlebug – including one I found buzzing frantically in my filing cabinet, and another in my sandal.

This weekend, I saw the cockchafer's even more handsome relative, the rose chafer, which looks like it has been spray painted with iridescent emerald green; while David and Kate caught a large and fearsome-looking black beetle with prominent horns, which they photographed before releasing. It turned out to be a lesser stag beetle: not quite as impressive as its better-known cousin, but still a good find.

One result of leaving the doors and windows of my office wide open during the hot weather is that insects are not the only creatures to come inside. One morning I heard a familiar series of thin, high-pitched 'seep seep' calls from the elderflower bush outside; moments later, a tiny bird flew through the doors and started to flutter against the glass, temporarily unable to escape. I soon managed to cup my hands around what was a newly fledged juvenile long-tailed tit – surely one of the cutest birds imaginable – and, taking it outside, reunited it with its family.

On the final night of the Bank Holiday weekend we all gathered outside to look out for the International Space Station, which was due to pass overhead at 10.55pm. Using their phone tracker apps, David and Suzanne managed to 'see' the approach, even though the spacecraft was not only out of sight, but actually over the horizon, so the app was 'looking' through the Earth.

Then, bang on time, a bright star-like dot appeared in the still just-about-light western sky and passed rapidly over our heads. To someone like me, who came of age at the height of the Space Age, and has a slightly unhealthy obsession with the moon landings, seeing this spacecraft was truly awe-inspiring. I also reflected on what it must be like to be in orbit two hundred miles above the surface of the Earth, travelling through space at roughly seventeen thousand miles per hour, or almost five miles every second. Now that really *is* self-isolation.

* * *

There are, of course, downsides to this glorious spell of unbroken fine weather. Last week, news broke of a major heathland fire in Dorset, at Wareham Forest. Described by the local fire and rescue service as 'one of the most devastating forest and heath fires in Dorset in living memory', it destroyed almost six hundred acres of prime bird and wildlife land.

The fire was caused, inevitably, by human carelessness: afterwards, discarded barbecues, campfires and glass bottles were found littered all over the site. The hot weather and very dry conditions soon allowed the fire to spread out of control, and although fortunately no one was hurt, the area will take decades to recover. In the meantime, specialised creatures such as Dartford warbler, sand lizard and smooth snake must squeeze into ever tinier patches of this precious habitat. I also reflected on the sadness and devastation felt by those people for whom Wareham Forest is *their* local patch.

A few days later, there was another disastrous fire, at Hatfield Moors, part of the Humberhead Peatlands National Nature Reserve in South Yorkshire. This will have a particularly negative effect on nesting nightjars, which would have been slap-bang in the middle of their breeding season. And to cap it all, three people were arrested at an illegal rave at a nature reserve near Leeds city centre.

When the Prime Minister refers, as he often does, to 'good old British common sense' (as if all foreigners somehow lack this capacity), I can't help reflecting on the idiots who think that having a barbecue on a nature reserve during the driest spring in living memory is a good idea. Then again, last weekend also saw reports of huge crowds on England's beaches, and, with all public toilets closed, people openly defecating in the streets. So much for 'good old British common sense', eh, Mr Johnson?

As the warm weather has continued into the last week of May – with no real prospect of change – the vegetation along the rhyne behind my home is beginning to take over the water, as sturdy reeds encroach from either bank. The duckweed is now so abundant it covers virtually the whole surface, reminding me of the dark folk-tale of Jenny Greenteeth. She is a shadowy spectre who is supposed to lurk beneath the layer of duckweed, from where she tempts young children to step onto the solid-looking vegetation and sink through to their deaths.

That evening, the only living presence in the rhyne was a pair of moorhens, feeding quietly by dipping their red-and-yellow bills into the water in search of aquatic invertebrates and, indeed, the duckweed itself, for moorhens have a fairly catholic diet. I sneaked along to just above where they were feeding and peered through the reeds, to try to get a closer look at them, but they had already retreated out of sight, as they so often do.

The name moorhen is rather misleading: they are not a bird of the high, open uplands of Dartmoor or Exmoor; instead, 'moor' is a corruption of 'mere', a small area of standing water, a habitat for which the moorhen is usually the only waterbird to be found.

A year or so ago, when walking with Rosie along this same stretch of water, I saw another, even more curious and elusive creature: a small mammal doggy-paddling across the duckweed. My first reaction was that it must be a water vole, but it was far too small; my next thought was that it was a shrew, but if so, why was it so dark in shade, and what on earth was it doing swimming across the rhyne? Then the penny finally dropped: it was, of course, a water shrew, which has the strange distinction of being Britain's only venomous mammal, whose saliva can incapacitate its prey.

As I was recalling this special moment, another shy and overlooked creature flew over me: a flash of bright cherry-red, bookended by a black head and white rump: my first bullfinch of the year here. I do frequently hear them in our garden, uttering that soft, melancholy 'piu piu' sound; but they rarely show themselves for more than a few seconds.

I then remembered that, normally, if I were cycling along this road towards the Mendips on a fine late spring evening, I would be on my way to the Wheatsheaf Inn, three miles to the north in the village of Stone Allerton, to meet my friends Dave and Nick.

The Wheatsheaf would have been rated highly by George Orwell, for whom the search for the perfect pub was a lifelong obsession. With its flagstone floors, sturdy wooden tables, real ales and Somerset ciders, no-nonsense home-cooked food, and friendly staff, the Wheatsheaf finds the perfect balance between tradition and innovation. I miss going there more than almost anything else that lockdown has deprived me of.

I do wonder, at moments like this, whether life can ever be the same, even if we do eventually get 'back to normal'. As Grace Dent wrote in the *Guardian*, 'Things will move on, but I know I will never take my right to loiter and languish, to laugh with friends or to drink lager from a pint glass for granted ever again.'

Those of us living in rural Britain are, in many ways, insulated from the current crisis; though yesterday it was reported that our local hospital, a few miles down the road at Weston-super-Mare, has closed its doors to new admissions because it is overwhelmed by Covid-19 patients.

But an item on tonight's *Six O'Clock News* really brought home how fortunate we are. Clive Myrie presented one of the most sensitive, beautifully balanced and moving news reports I have ever seen: from the Royal London Hospital, in the heart of the capital. I was moved to tears by the remarkable hard work and compassion of the NHS frontline staff, and the sheer horror of what has been happening there. It was a timely reminder, if we needed it, that the crisis is far from over, and that lifting lockdown restrictions could easily come back to bite us, as indeed top scientists are beginning to warn.

Shaking off these melancholy thoughts, I carried on around the loop, watching low-flying swallows hoovering up insects from fields that, following the first cut of silage, have been shorn to the length of a corporal's crew-cut. Apart from their companionable twittering, all is quiet again, now that the tractors which have been racing up and down these narrow lanes have been put away, at least until the next silage cut in a few weeks' time. That, however, does depend on the coming of the midsummer rains, which for the moment at least, look less and less likely to arrive in this increasingly parched and yellowing landscape.

*   *   *

Elsewhere, a row has kicked off in New York, following a tense encounter between two visitors to Central Park: a dog-walker and a birder. It started when the birder politely asked the dog-walker to keep her animal on the lead, as instructed by prominent signs around the area known as The Ramble, a well-known hotspot for migrant birds passing north in spring. Instead of politely complying, the dog-walker then took exception to the birder filming her on his smartphone, and threatened to call the cops.

What I have not mentioned, but is clearly relevant to the whole incident, is that the dog-walker was a white woman and the birder a black man. And that she then called the police, because 'there's an African-American man threatening my life!'

Afterwards, social media predictably went berserk, with considerable justification: as many people pointed out, this was a classic example of the insidious undercurrent of racism that regards any black person – especially a man – as a threat. Had the birder not walked away; had the cops actually come to investigate; we might perhaps now be reading about the birder being arrested, or even shot and killed. On the other hand, it is clear from the video that the woman did herself feel threatened – however unjustifiably – and was in a state of blind panic. Perhaps this says more about the sad state of American race-relations than anything else.

The American Birding Association was quick off the mark with a clear and considered condemnation of what had happened:

We believe that all birders should be able to participate in their hobby free of harassment and bigotry, and we acknowledge that this is frequently not the case for birders of color... Inclusion and equity are core ABA values; fear and intimidation should never be part of birding culture. Access to outdoor spaces without fear must be a right for all who seek to enjoy and protect wild birds.

The irony is that all the poor birder was doing was trying to enjoy an innocent pastime in the heart of one of the world's busiest cities, at a time of global uncertainty. As my virtual Twitter friend Frank Izaguirre noted, in a still, small voice of calm: 'College cancelled. Travel cancelled. Sports cancelled. Spring migration NOT CANCELLED. Thank God for birds because this shit is scary and stressful. Hope everyone is safe.'

A week later, the whole of America went into meltdown because of the killing by a white policeman of a black man, George Floyd, making the incident in Central Park seem trivial by comparison. Yet they are both part of the same global problem of inequality between the races, and reveal that any talk about a kinder, more understanding world following the coronavirus crisis is at the very least premature, and may never come to pass.

* * *

On 28th May, this bizarre spring turned – almost overnight – into full-on summer. The morning dawned, as they do every day at present, sunny and warm, and during the day the heat then increased inexorably until – even with all the doors and windows in my office wide open – it became too hot to work there, with my thermometer topping the thirty-degree mark.

Emerging for a brief respite, I noticed a small butterfly moving purposefully along the strip of lavender beside the lawn. At first, I thought it was a common blue, but a closer look revealed that this was that species' smaller and more exquisite cousin, the brown argus.

Ironically, given the small size of this butterfly, it is named after an all-seeing giant with one hundred eyes, from Ancient Greek mythology; in this case, because the underwings are dotted with black spots, each surrounded by a ring of white, like eyes. Although this is technically one of the 'blues', the brown argus is, as its name suggests, a deep chocolate-brown shade above, with its upperwings bordered with orange smudges, and rimmed with white, giving it a pleasingly neat appearance.

Like so many butterflies I have seen in the garden this year, the brown argus was several weeks earlier than I would usually expect – they generally appear in mid-June – so I decided to take a stroll to the end of our garden to see if the commonest of our summer butterflies had emerged yet. Sure enough, almost immediately I came across a meadow brown fluttering along by the fence – a full ten days earlier than the previous earliest record here.

Meadow brown is the default butterfly of high summer – on the wing throughout June and July, with a few faded, tatty individuals hanging on into August and early September. It is actually Britain's commonest and most widespread butterfly, but because it isn't really much to look at – mid-brown with paler brown and orange patches and false eyes on the forewings – it is often overlooked in favour of its more colourful relatives.

The appearance of these midsummer butterflies (and I saw even more meadow browns that weekend, when Daisy and I took a cycle ride around our neighbouring villages) suggested that the 'June gap' really had come earlier this year – in mid-May. And this was far from the only sign of summer. Our lawn has hardly needed mowing this year because of the lack of rain; and I have also noticed that, with the increasingly hot days, the tell-tale strips of brown where pipes have been laid or dug up in the past are again beginning to appear, just as they did during the past two drought summers. But in those years, it took until July for the dry weather to have this effect: this year, it has happened in May.

(The Met Office later confirmed that this was the driest and sunniest spring across the UK since records began, with just ten millimetres – less than half an inch – of rainfall in the whole of the month of May, and an average of six-and-a-half days of sunshine per day for the whole three-month period. Indeed, this was not just the sunniest May, but the sunniest *month* ever – beating the previous record held by June 1957.)

Much as I love this weather, my enjoyment is tempered by the fear that this is a permanent shift, due to the climate crisis; one which augurs ill for nature and humanity alike.

On a lighter note, this was also the anniversary of the day, three years ago, when we acquired Rosie. At the time, I was the most reluctant of the five of us to embark on the huge responsibility – and expense – of owning a dog. But after constant pressure from Suzanne and the children, I finally gave in, despite being sceptical about her claim that it would be good for me – 'when you're all alone in your office, you'll have company…' – and also encourage the children to get out more.

But – and this is a phrase that rarely leaves my lips – I have to admit that I was wrong. Rosie has brought untold joy into our lives, and she really has improved our physical, mental and emotional health.

*   *   *

Week Ten

During lockdown, my old friend Marek Borkowski has been periodically sending informative emails from his home deep in the Biebrza Marshes in north-eastern Poland, where he lives with his wife, Hania, and their two younger children (their two older ones are studying abroad). Normally at this time of year, Marek and Hania would be incredibly busy leading their famous bird tours, but this year they are confined to home.

I first got to know Marek more than twenty years ago, when we filmed an episode of the TV series *Birding with Bill Oddie* in Poland. Even now, people regularly tell me that this is their favourite of all the programmes we made. Watching it again recently, the sheer number of wonderful birds – nesting white storks, singing aquatic warblers and a dazzling colony of white-winged black terns among them – made me realise just how rich and special the birdlife of Eastern Europe is, especially compared with our own dwindling and depleted avifauna.

Since then, we have met Marek and his family each year at the annual Birdfair in Rutland, and he has occasionally stayed with us here in Somerset. But we had never – despite numerous invitations – been back to Poland, at least until last April, when, with Charlie away on a school trip, Suzanne, George, Daisy and I visited the family as their guests.

It was a wonderful few days: statuesque pairs of white storks on nesting-platforms in every village; pygmy owl and nutcracker in the woods near his home; no fewer than eight different kinds of woodpeckers, including the elusive three-toed; and, most memorably, a dozen different mammals. These included elk, European bison and – the undoubted highlight of the trip – a lone wolf, spotted by Suzanne as it wandered across a dirt track on the outskirts of a village, in the middle of the day.

So, I have really enjoyed the long, rambling yet always fascinating messages Marek has been sending out to his friends and colleagues all over the world, entitled 'e-birding in the time of Plague'. These began, in early May, with a stunning photo of a male capercaillie – the world's largest grouse – perched in a tree at dawn. This was accompanied by Marek's wonderfully idiosyncratic description of the experience:

> *Imagine pitch dark morning.*
> *Temperature minus three.*
> *Absolutely quiet.*
>
> *The only noise…*
> *A corking sound of a male Capercaillie.*
>
> *He is black, but when turning the black is shining purple, green, blue and nearly every tint in between… bringing to mind a petrol blur on a surface of a puddle…*

Marek then went on to inform us about the major fire that had recently broken out on his beloved marshes, due to the unprecedented dry weather. He pointed out that fires were not necessarily a disaster there – they have always been part of the natural regeneration process. But what, ironically, made this one so much worse was the knee-jerk response of the authorities, 'whose reaction to "fire" is always bad (a bit like a "wolf") and needs to be fought (killed)'.

As a result, what Marek described as 'herds' of heavy fire trucks and hundreds of firefighters invaded the marsh, slap-bang in the middle of the breeding season. The damage, he told us, will take years to repair. Very sad.

To try to counter the gloom, Marek's later dispatches were lighter in tone: a black woodpecker bouncing along in flight, against a bright blue sky; followed by a very different view of a white stork perched on a fence post, surrounded by snow. It seems that while spring has been edging closer and closer to summer here in Somerset, in Poland winter has returned with a vengeance.

* * *

The final weekend of May, and we have been told that many lockdown restrictions are now to be lifted, including allowing gatherings of up to six people, visits to friends and relatives, and even barbecues.

Predictably, hundreds of thousands of my fellow coun-trymen and women have taken this opportunity to fur-ther demonstrate their 'good old British common sense' by flocking to parks, beaches and other visitor hotspots. Hordes of people queued for hours to get into car parks, crammed cheek-by-jowl onto the grass, sand or shingle, and in every possible way flouted the still barely extant code of conduct. They also left vast amounts of litter behind them.

It seems to me that the tragedy of the whole Cummings scandal is not its political fallout, but the disastrous effect it has had on public trust; so that the Health Secretary Matt

Hancock's feeble exhortations to us to abide by our 'civic duty' are likely to go not just unheeded, but flouted with impunity.

And yes, I know how lucky I am: to be in lockdown with so many of my loved ones, in a large home with an even larger garden, and with a place I can enjoy my favourite recreation of birding so close to where I live. But I'm not the only one concerned that the sudden lifting of so many lockdown restrictions has taken place only to divert attention away from the Cummings scandal. It is the modern equivalent (and the PM will, with his much-trumpeted classical education, recognise the reference) to the Romans' 'bread and circuses': policies which pacify the people in the short term, but leave us open to future perils as yet unknown.

It has been said – and it may well be true – that as a nation we will emerge from this crisis in a better place: not, perhaps, financially or economically, but in terms of a greater kindness and understanding of one another, even when we might choose to disagree.

Again, it's a lovely thought, but sadly it may not apply to everyone. On the last Saturday of the month I awoke and checked my email, to find a message from the conservationist and blogger Mark Avery. I have known Mark, on and off, for over forty years, ever since we first met at Dungeness Bird Observatory in the late 1970s. To be honest, we didn't get on very well then, but years later, once we had both established careers in the field of wildlife and conservation – me in natural history television and he at the RSPB – we formed a close professional friendship.

Week Ten

I admire Mark hugely for his campaigning skills, and his dogged refusal to allow vested interests – whether in farming or field 'sports' – to triumph. Along with our mutual friend Chris Packham, Mark is probably loathed by more people in the grouse- and pheasant-shooting industries than anyone. Yet he continues to campaign, undaunted, for what he believes is right.

His message brought me up with a start: my appearance on the *Today* programme, in which I had mentioned that birds of prey are currently 'doing incredibly well in Britain' had been used in a blog by a new organisation called the Campaign for Protection of Moorland Communities (or C4PMC) to condemn campaigners against the illegal persecution of raptors such as the hen harrier on grouse moors.

At first, I found this blatant misrepresentation of my views mildly amusing; then, as I re-read the posting, I became more and more angry. I was somewhat mollified when Mark posted his response, a witty and pointed retort to their ridiculous and unfounded claims; and when I discovered that this organisation did not have a strong social media presence.

Nevertheless, it rankled for the rest of the weekend, especially as the purpose of my radio interview was to encourage people confined to their homes in cities to look out for raptors in the skies above. As someone who has moved from the city to the countryside, it always amuses me that these self-appointed champions of rural life complain about how their urban counterparts don't understand why they should have the right to illegally kill hen harriers; yet show no appreciation that city-dwellers value and cherish wildlife at least as much – if not more than – they do. And as we constantly remind them, they are defending the indefensible, and are on the wrong side of history.

Week Eleven

1st–7th June

Marbled White

Perhaps for the first time, the Met Office have got it right when they claim that 1st June marks the start of summer; the day dawned – as they have done for the past few weeks – bright, sunny and warm, as I soon discovered when I realised I had overdressed for my early-morning bike ride.

Spring is still clinging on, at least judging by the level of birdsong in my garden and around the moor: chiffchaffs and blackcaps carolling me as I awake, and sedge warblers manically launching themselves into the air from the thick hawthorn hedgerows around the loop.

But summer is most definitely starting to take over. Later that morning, an emperor dragonfly – one of our largest insects – cruised around in the sunshine above our garden deckchairs, grabbing tiny insects as it went; while in the heat of the afternoon, Suzanne and I sat outside and watched as a buzzard continually launched itself up into the air, stalled and, folding its wings, plummeted back downwards. 'It must be having so much fun,' said Suzanne, and – anthropomorphism aside – she was surely right.

When we launched *Springwatch*, back in 2005, we got a lot of grief from people who claimed – with some justification given the Met Office's view that spring covers the months of March, April and May – that by the time we entered our second of three weeks' broadcasts, spring was actually over, and that we finished the programme just a week before Midsummer's Day. I countered this by pointing out that the 'mid' in 'midsummer' comes from the same root as 'midwife' – and means 'with', rather than 'in the middle of'. Thus, Midsummer's Day marks the beginning, not the middle, of the summer season.

All that now seems rather academic, given my experience yesterday, when the thermometer reached a sweltering twenty-six degrees, and I headed down to Huntspill Sluice to search for summer butterflies. I found them: meadow browns were feeding on virtually every bramble flower and cow parsley bloom, the all-dark males and more varied females, with their orange wing-patches, competing for nectar.

In half an hour spent wandering around the long grass above the sluice itself, I also came across black-tailed skimmer dragonflies, blue-tailed damselflies, and more butterflies, including a single common blue, brown argus and small heath – looking like a paler, miniature version of the meadow brown.

Best of all, I stumbled across two large skippers, which despite their name are one of the smallest of our butterflies, and have a moth-like appearance, resting with their triangular wings closed. These were the first large skippers I've ever seen here, and the twenty-second species in all – meaning that I have seen more than one-third of all Britain's fifty-nine species of butterfly on this modest coastal patch.

The following Sunday I went down again, in cooler and breezier but still sunny weather, on my way to getting my car cleaned. As well as the plentiful meadow browns, I also caught sight of several marbled whites. With their piebald coloration, they are one of my favourite butterflies. That day it was also confirmed that this has indeed been one of the earliest – if not *the* earliest – butterfly seasons on record, with many species emerging in good numbers at least a month earlier than usual.

Butterfly guru Matthew Oates, who has been assiduously noting the first dates he sees butterflies each year since 1971, has seen no fewer than ten species (including the marbled white) earlier than in any year since then. Indeed, only three of our native species have yet to appear.

Matthew thinks that one reason butterflies are doing so well this year is that this is the third year in a row with warmer and sunnier weather than usual. However, he too is concerned that although they are doing well in the short term, such rapid change may soon turn out to be a problem, especially for our scarcer and more specialised species, as they may go out of synch with their foodplants – a similar issue that faces migrant birds. Once again, this fine weather has a nasty sting in its tail.

*   *   *

My friend Graham, who lives overlooking the River Thames at Limehouse in east London, has emailed me with an extraordinary picture, taken from his balcony, of a grey seal happily popping up with a fish in its jaws, before swimming upstream.

Graham also mentioned how much lockdown has changed the way he looks at the familiar landmarks of the capital. 'You can see them as they should be,' he told me, 'instead of being obscured by people and traffic.' The river, too, is unusually calm and clear, because of the almost complete lack of boats travelling up and down. But on hot, sunny days, everything changes. 'When everyone who lives near the river decides to take their daily exercise, the only place they can go is the towpath – and it's complete chaos!' Once again, I reflected on the contrast with my daily perambulations around the loop, where I rarely meet anyone.

Last night, just before sunset, David, Kate and I decided to take Rosie for a walk. But, as usual, she took a different view. I have never known a dog who does not respond positively to the word 'walkies', especially if uttered with enthusiasm. Rosie, however, is the exact opposite. Not only does she cower when we produce her lead and harness, but she seems to sense when we have planned a walk, even when we innocently call her indoors.

As a result, we ended up taking the walk as a dogless trio. Even at quarter-past-nine the sun had not quite sunk below the distant shape of Brent Knoll, and when we returned, just in time to watch the *Ten O'Clock News*, there was still enough light in the sky for it not to feel even close to darkness.

This is by far the longest time David and Kate, who were both brought up in London, have ever spent in the countryside. I asked them whether they have been feeling closer to nature as a result. Kate definitely does: their daily runs and ever-more ambitious bike rides have opened her eyes to the beauty of this little corner of the English countryside. Typically, David takes a contrary view, pointing out the artificial nature of this farmed landscape. I see his point, but I wonder whether, after they have completed their planned cycle-ride from Land's End to John O'Groats in the summer holidays, he will change his mind.

Halfway along the southern border of the moor, we stopped to listen to a sedge warbler. Unusually for such a showy bird, it was buried deep inside a thick hedgerow, and although we stood and listened to it singing for several minutes, it resolutely refused to show itself. Maybe it had heard David's disparaging remarks about its spring and summer home.

For me, the sedge warbler's song has been the soundtrack of lockdown. Unlike the reed warbler, which strings together a rather uninspiring series of repeated notes, the sedge warbler's song is full of vim and vigour. There are two main components: a grating, irregularly rhythmic base overlaid with fluty whistles, almost as if two completely different birds are singing.

Fundamentally, the sounds made by these two closely related migrants reflect their personalities: the reed warbler is shy and retiring with a rather predictable song to match;

the sedge warbler, which sounds like a jazz musician riffing on a theme, is far more outgoing and extrovert in nature. Not tonight, though.

*   *   *

One of the joys of Twitter in lockdown has been the wit and imagination of some of my fellow-users. As well as the 'James Bond wildlife', we have had the 'change one letter of a bird's name' (smew to stew, garden warbler to warden warbler, yellow wagtail to mellow wagtail, and so on), and various other rather silly but nevertheless entertaining threads.

Then there have been the virtual 'World Cups': including the Seabird World Cup (won, predictably, by the puffin) and, more recently, the Moth World Cup. Organised by young naturalist James Beaumont, whose Twitter profile tells me that he is 'Proud to be autistic!', it was won – to my delight – by the hummingbird hawkmoth, the creature I celebrated in the title of my last book about the wildlife of my parish, *Wild Hares and Hummingbirds*.

James is just one of a new generation of youngsters who have seized the opportunities provided by social media to talk about their connection with and love of the natural world. Not all that long ago, many well-respected commentators were lamenting the 'death of the naturalist', suggesting that, because young people had become so disconnected from the outdoor experience, when the older generation die off there would be nobody to replace them.

But, in less than a decade, that situation has completely reversed. Young people are at least as engaged with the natural world as my generation was – arguably even more so. Their secret weapon is social media, which has helped them find others who share their passion and create a real and virtual community of like-minded souls.

This came to a head a week or two ago, when Dara McAnulty, a sixteen-year-old from Northern Ireland, read his new book, *Diary of a Young Naturalist*, on Radio 4's Book of the Week slot. I have followed Dara's incredible journey from a shy, bullied teenager to a far more confident one with fascination, and it has been great to witness his success.

*   *   *

This morning dawned cool and cloudy, following the unexpected sound of rain falling on our bedroom roof during the night. I marked this change in the weather by finally finishing off the last pot of home-made marmalade given to me by the woman my children call 'the marmalady': our friend June.

I first met June soon after I moved down here, when I gave a talk to an RSPB group just over the border in Dorset. She greeted me as if I were an old friend, which I now am, and said very kind words about my books. The next time we met she asked me if I might be able to send a book to a friend of hers who had recently been bereaved; I did, and a few weeks later she turned up at my home with half-a-dozen pots of marmalade to say thank you.

For the next decade, she continued her annual visit, and at every breakfast, as I spread the marmalade onto my toast, I am reminded of her kindness – and, by association, of mine, now more than repaid. We have, on occasions, gone out birding together; the last time we did so she photographed a striking black-and-yellow dragonfly, which turned out to be a female scarce chaser, the first I had ever seen in Somerset.

Last year, June apologetically informed me that this would be the last batch of the orange condiment; and so now, a year later, the morning ritual is over. But as a result of that chance encounter, we have become friends, and at a time like this, when small gestures are so significant, we should all value human connections such as this.

*   *   *

As we have all had more time to think and reflect during lockdown, it's not surprising that several influential figures are calling for a permanent change in our relationship with the natural world after all this is over. In an online talk back in April, entitled 'Pandemics, Wildlife and Intensive Animal Farming', the primatologist Dr Jane Goodall called for a global ban on the trade in wildlife – which it appears was probably responsible for the mess we now find ourselves in, if the theory that Covid-19 started in a Wuhan animal market is correct. Even more significantly, she also asked for an end to this 'destructive and greedy period of human history', now officially defined as the Anthropocene, in which the influence of humanity on life on our planet has come to dominate all other natural processes.

This week, Prince Charles lent his voice to support Dr Goodall's arguments, calling for us to put nature 'at the centre of everything we do... The more we erode the natural world, the more we destroy biodiversity, the more we expose ourselves to this kind of danger.' I have always been rather ambivalent about the heir to the throne, but when he takes the long view about our relationship with the natural world, I must say I agree with him.

Meanwhile, the news has been getting grimmer and grimmer. The days when, back in March, the government's Chief Scientific Officer suggested that if we could limit the death toll from Covid-19 to twenty thousand, this would be 'a good outcome', seem to come from another time. It seems that the total is now close to twice that – not counting those thousands of poor souls who died without being diagnosed with the disease, or who have died from other causes because they were unable to get the treatment they so desperately needed.

For a few months towards the end of last year and the start of this, I was undergoing tests for a possible problem with my prostate. It came as a shock to realise that, as I approached my sixtieth birthday, I was now in the age-range considered to be most vulnerable to the potentially fatal disease of prostate cancer. Probes and blood tests were followed by a biopsy, then an MRI scan, followed by more blood tests, and finally both CT and bone scans. Each time there was good and slightly less good news: 'We haven't found anything wrong with your prostate but there's something strange in your pelvis/skull/etc.'

Finally, at the beginning of March, I got the all-clear, an enormous relief at the time; but even more so in the light of what happened just two weeks later. So, I have enormous sympathy for those unfortunate people who are still being tested for various cancers, and other conditions, and have had to wait in limbo because of the current crisis.

*   *   *

One of the unexpected bonuses of the lockdown is the huge rise in the number of events being held online. Whereas before I would have been reluctant to travel to see a talk, now I can simply attend from my office, as I did yesterday.

The speaker was zoologist and campaigner Alasdair Cameron (who, for reasons best known to himself, goes under the Twitter handle @TheMushyPea). He was talking about a subject much in the news at the moment – rewilding – not on a national or regional scale, but at the local level.

A few years ago, Alasdair bought two fields – one wet, the other fairly dry – on the edge of the village of Godney in the heart of the Somerset Levels. After making a few changes – digging a scrape to create a permanent waterbody, and introducing some (very cute) harvest mice – he then basically left the land alone to see what happened.

Against the grain of the prevailing conservation philosophy, which calls for constant habitat management, this was a huge success. As he told us, things don't change overnight, but just five years on, the landscape has nevertheless been transformed. There are thistles, which attract countless butterflies in summer and huge flocks of goldfinches in autumn; flooded areas, attracting flocks of lapwings and plenty of snipe in winter; huge molehills; and a range of other creatures, including roe deer and hares, grass snakes and adders, and vast numbers of invertebrates, notably spiders, dragonflies and damselflies.

This may be small-scale, but it's also very easy and inexpensive to do, and as Alasdair pointed out, 'If every landowner set aside a small plot of land, which attracted as much wildlife as I get here, things would be an awful lot better.' He was very encouraging to the many attendees who had similar ambitions: 'Don't be afraid – just do it!'

The session ended with what he called 'a message of hope': of how, if this kind of small-scale rewilding became the norm, there would be so many benefits – not just for the wildlife, but for our culture and economy, too.

I went away feeling both uplifted and excited for the future of the landscape that means so much to me. I have long had an ambition to buy the field immediately behind the bottom of my garden from my farmer neighbour Rick, and turn a grassy area used for grazing sheep and growing silage into my very own local nature reserve. I would then – of course – write a book about it, entitled *Field of Dreams*. Alasdair's talk fired me up and made me think more seriously about turning my own dream into reality.

Later the same day I chaired a consultation session on the proposed GCSE in Natural History. A few years ago, my friend and former colleague, the conservationist and author Mary Colwell, told me of her idea to create such a qualification; and I must say I was pretty sceptical. It seemed to me that the educational establishment – both the government and schools – would not welcome this addition to an already crowded curriculum.

But I failed to reckon on Mary's tenacity, and her sheer unwillingness to accept the word 'no' for an answer. So here we were, with two hundred and fifty of the most influential people in the worlds of education and the environment, launching a nationwide consultation to help shape a qualification that could, in turn, help shape the future of the world we live in.

Again, I ended the session on a huge high – a tribute to the way that when those of us with a passion for the natural world act together, we can achieve incredible things.

*   *   *

As June progressed, and the winds strengthened, birdsong in our garden dropped to the bare minimum. Yet one little bird has continued to sing and proclaim its territory every single day since it first arrived back from North Africa on 19th March, four days before lockdown began.

Each morning, as I commute the fifty feet or so from my home to my garden office, I have heard it calling out its name from the scrubby area of cider apple trees, ash saplings and brambles beyond the dog fence at the bottom of our garden.

'Chiff-chiff-chaff-chiff-chaff,' it sings, making me wonder why some female birds, such as the nightingale or marsh warbler, expect their potential mates to be virtuoso songsters. Males of these two species create complex (and ever-changing) patterns of hundreds of different notes and phrases – in the case of the marsh warbler, faithfully mimicking the songs and calls of dozens of species from both

Europe and Africa – into what sounds to us like a particularly complex jazz improvisation session.

In complete contrast, male chiffchaffs sing the same two notes, in a very limited combination, time after time. Presumably, female chiffchaffs are either simply less fussy than the females of other species, or are looking for something else in their mates: skill at pumping their tail up-and-down, perhaps, or a nice shade of olive-green in their plumage?

Despite their rather drab appearance and monotonous song, I have a soft spot for chiffchaffs: they really are the first sign that spring is about to kick off – and, especially this year, have been a welcome accompaniment as I sit and work here, day after day.

On the first Sunday of June, summer returned – well, a pale shadow of what we had enjoyed in May, but still sunny and fairly warm. I took the opportunity to do my second and final BTO Breeding Bird Survey of the season, a transect through my village from our home.

I was pleasantly surprised by how many birds I saw and heard: twenty-one species, many of which were still singing; in the case of the robin and wren, some five months after they first began to do so. Highlights included a couple of house martins calling above the timber yard up the road, a calling green woodpecker, singing goldcrest and a raven.

In Little Moor Lane, I cycled past a couple of raucous peacocks (sadly, they don't count for the survey), and in the far field came across a flock of two hundred rooks, many of them adults feeding well-grown young. There are plenty of rookeries around these parts – including one, aptly, in the nearby village of Rooksbridge on the A38 – but they are not popular with the local farmers, who claim that they kill their lambs. As the rooks called to one another, I was reminded of Dominic Couzens' way of deciding whether the cawing sound you are hearing is a rook or a crow: 'A rook sounds like a crow who's been on an anger-management course.' Brilliant – and it works.

On my way back, via the village stores to collect the Sunday papers, I was overtaken by little flocks of cyclists, all togged up in their Lycra and travelling at a much faster rate than me. As they sped past, though, I was struck by how virtually all of them greeted me cheerily; have manners really improved due to lockdown, or were they always so polite?

Week Twelve
8th–14th June

Nightjar

The birds have still been singing on my early-morning bike ride around the loop, even when, as frequently happened this week, strange wet stuff was falling out of the sky like bathwater slowly leaking through the ceiling. Worse is, apparently, still to come, with downpours forecast for Friday and Saturday. Still, as everyone rather tediously points out, our gardens need it.

Over the past few days, I have heard all the usual warblers, belting out their songs from the hidden depths of the hawthorn hedgerows. Chiffchaff, whitethroat, lesser whitethroat, reed and sedge warblers all either can or do have a second brood of chicks; and given the mainly very fine weather this spring I suspect they are mostly doing so.

Skylarks with Rosie

I often use my time on the bike to ponder, and this morning I considered the odd differences between the birdlife of my garden and that of the moor. Separated by less than fifty yards – so close that I can sometimes hear one bird singing in the garden when I am on the moor, and vice versa – there are nevertheless several species that are either common in one and very scarce or simply absent from the other.

Take our commonest summer migrant, the willow warbler. When we first moved here, in 2006, I would hear willow warblers every year, for a week or so each spring, from the scrubby area at the end of our garden. These were not breeding birds, but just passing through on their way north. Nowadays, several years might go by without me hearing that familiar, plaintive song; this year I heard just one, on a typical date of 9th April. Yet I have never heard or seen one on or around the moor.

Conversely, while reed warbler and common white-throat have turned up in my garden on two or three occasions, they are common on the moor; as are sedge warblers, which I have never seen or heard in the garden. Chats, too, are regular visitors to the moor, with stonechat, whinchat, wheatear and even a single redstart turning up; but I am still yet to add any of these to my 'garden list'.

Habitat is clearly the key: the garden provides plenty of trees and bushes, which suit some species, while the moor is mostly grassland and hedgerows, dotted with clumps of reeds. That still doesn't explain why that reed warbler was belting out its song from a lilac next to our house, a year after we moved down here.

The current garden sounds come mainly from the 'Three Gs': greenfinch, goldfinch and goldcrest. I came out of the back door the other day to the sight and sound of a greenfinch in full-on display flight: going as slowly as it dares, and trembling its outstretched wings while uttering that distinctive, wheezy song. Goldfinches frequently fly overhead, their tinkling sounds falling down like sonic glitter; while, unseen inside the densest foliage, a goldcrest constantly sings its rhythmic, high-pitched song.

Goldcrests are Britain's smallest bird – a shade smaller than the wren, but just half its weight. A fully-grown adult goldcrest weighs just five grams: the same as a twenty pence coin or a single A4 sheet of paper. Think about it: one of those quarter-pound bags of sweets we used to buy on the way home from school would – if the sweets were swapped for goldcrests – contain more than twenty of these tiny creatures. Indeed, a single sherbet lemon weighs almost twice as much as a goldcrest. Simply amazing.

This week has also seen the return of the magpies: four of them this time – all well-grown youngsters – whose rattling, machine-gun calls are a constant soundtrack as I work, a call that led to this extrovert corvid being dubbed 'chatterpie' or 'chatternag'.

The other day, all four of them discovered the two reclining chairs Suzanne had recently bought from the local garden centre. For a few minutes, they moved adroitly in and out of the frames, up and down, round and round, as if completing a particularly complex obstacle course. Then one was distracted by the sight of something else, called loudly to his companions, and they all flew up into the nearby elder. They are also frequently perching on the children's (largely unused) trampoline; but they do need to watch out, as Rosie often seeks shelter from the sun beneath it. If she hears a curious magpie pecking at the surface, she may well try to catch the culprit. She might even succeed.

*   *   *

Another new – and not quite so welcome – visitor is a family of rats, which have made their home in a hole beneath our garden bird feeder. While doing the washing-up, I have been treated to views of them: at first, one shyly poking a long, twitching nose out of its lair, then climbing up onto our stone bird bath and finally, brazenly, the whole family – mum, dad and three smaller youngsters – shinning up the metal pole to feed on our birdseed.

I struggle to like rats. And I'm in good company. David Attenborough once confessed, 'I really, really hate rats. I've handled deadly spiders, snakes and scorpions without batting an eyelid, but if I see a rat I'll be the first to run…' I've long felt the same: and yet I have a sneaking admiration for them, too. They are what scientists call 'neophobic': deeply suspicious of any changes in their immediate environment – and that is the key to their success. We once tried to rig a remote camera in a tunnel where rats were living: they didn't just give this suspicious device a wide berth, they took to their heels and vanished.

And although technically the brown rat is an 'alien invader' – having been accidentally brought here on ships sometime in the late seventeenth or early eighteenth century – they have surely by now earned their status as a pukka British mammal. So instead of calling the pest controllers, I have learned to tolerate this little family, and even enjoy watching their acrobatic antics. Maybe I'm getting soft in my old age.

Later that day, glancing out of my office window at the rain, I noticed that the elderflowers that until now have been in full bloom are starting to fade – and quickly, too. I have a soft spot for these creamy blossoms: my mother and I used to collect armfuls of them so she could make gallons of home-made wine, demijohns of which sat in our understairs cupboard, maturing to a deep shade of yellow, like ancient specimens of urine. I can't vouch for what it tasted like; I was too young to drink. But, somehow, I imagine it was pretty awful.

*   *   *

On the second Saturday of June, in place of our usual outings to the Avalon Marshes or the coast, Graeme and I decided to explore a site much closer to home. I had last crossed Tealham and Tadham Moors with Daisy back in May, on our bikes, when there was not much time or opportunity for birding. So, this time Graeme and I left our bikes at a friend's home just above the moors, and walked the three-mile circuit, giving us time to take in the birds at a more leisurely and, we hoped, productive pace.

I had left home twenty minutes early for a quick circuit of the loop. The day before, it had rained heavily, and the air was still muggy and breezy, so the swallows were hunting for tiny insects barely a foot or two above the grassy field behind my home. It felt like a promising morning, and although I didn't see much on my way round, I did hear several white-throats, the ubiquitous wrens, trilling their hearts out, and of course the usual chorus of skylarks that has accompanied me on every visit here since the lockdown began. One welcome sighting was a family of baby blue tits, their yellow cheeks setting them apart from their parents, skittering along the tops of the hedges as I cycled past.

Earlier that morning I had opened the moth trap I'd set the evening before. I was prepared, once again, to be disappointed: this spring, whenever I have set the trap overnight, I have only found a handful of moths the next morning – and all those have been the usual 'little brown jobs' I struggle so hard to identify.

But this morning, I hit the jackpot at last. There were well over fifty moths, with some real crackers, including two different hawkmoths (the star prize for the moth enthusiast): a huge privet hawkmoth, with its intricately folded wings, and an elephant hawkmoth – a vision of cerise-pink and brownish-green, rather like the fuchsia plants on which they often perch.

These were accompanied by assorted buff and white ermines; a scorched wing, so-named after its appearance, which really does look as if the wings have been lightly singed with a blowtorch; a snout, a triangular-shaped moth with very prominent antennae; and a burnished brass, whose colour varies as you turn it towards the light, and reminds me of the iridescent gold and green of ancient brass ornaments.

But for me, the highlight was my favourite British insect; and, indeed, one of my very favourite creatures: a buff-tip. This moth looks exactly like a segment of birch twig covered with lichen, which has been neatly cut off at the end to reveal the new-growth wood. Then it opens its wings and you realise that it is not a twig at all. One of the great examples of how camouflage works in evolution.

When I was growing up, birdwatchers (as we called ourselves back then) were pretty dismissive of moths, but in recent years this has changed dramatically. Most birders now are fascinated not just with this large and complex group of insects, but also with butterflies, dragonflies and damselflies, too: anything with wings, really. Though I have yet to go as far as Erica McAlister and become obsessed with flies. Maybe that will come as I grow older and have more time and patience – a lot more time and patience, I suspect.

Back on Tealham Moor, Graeme and I set out as the sky threatened rain; but the clouds soon cleared and it turned into a beautiful morning. As we wandered down the lane, we heard first a lesser whitethroat, which as usual remained invisible, and then a common whitethroat, which proved more obliging, perching up on a tall stand of hogweed and delivering its distinctive song, before plunging down out of sight again, reminding me that the old folk-name for this species was 'nettle creeper'.

Singing meadow pipits and skylarks also accompanied us throughout the walk, their layers of sound receding into the distance. Then, a welcome surprise: a small, white egret landing by a herd of cattle in a damp, well-vegetated field. As we suspected, when we approached we could see the short, stubby bill, rounded wings and orange punk hairdo that marked it out as a cattle egret. Moments later, we noticed half-a-dozen more around the cattle, and then a score of others in a distant field.

Two decades ago, seeing two dozen cattle egrets anywhere in Britain would have led to a major twitch; now they have colonised the Avalon Marshes in force, and are the most numerous of the three species of egret found here. Graeme and I speculated, as we are wont to do, on the next species that might turn up here to breed: perhaps purple or squacco heron, spoonbill, or white stork?

We turned back for home, down a rutted drove, accompanied by meadow brown and small tortoiseshell butterflies – the latter thankfully a lot more numerous than they were earlier in the spring. Birds sprung up from the path a few yards in front of us: families of linnets, the youngsters so much duller than the adult males; goldfinches and reed buntings; and little groups of house martins and swifts passing low overhead. All in all, a great start to the weekend.

In some ways, Tealham and Tadham Moors are the future of this very special lowland landscape, showing that places don't have to be nature reserves to support plenty of varied and interesting wildlife. I resolved, there and then, to make this my newest 'local patch' next year. When I got home, I mentioned this to Suzanne – who suggested that was rather like changing which football team you support, from a mid-table struggler to a more high-ranking one. Good point, but I'm going to do it anyway. Like bird books, you can never have too many local patches.

Week Twelve

*   *   *

Less than a week to go before the summer solstice, which this year will fall on Saturday 20th June, and Suzanne and I went out on a date night. I say 'date night', but it was actually a trip up to the Mendip Hills to look for one of our most elusive and mysterious nocturnal creatures. Nor were we alone: we had invited Matthew, a local GP and Suzanne's business partner in their new and exciting venture to connect people with nature.

I had forgotten – as I always do – that Mendip has its own particular microclimate: less West Country, more Yorkshire Dales – or perhaps the Arctic Circle. The three of us shivered as we watched the sun set over Priddy Mineries, a nature reserve of grassland and shallow pools formed by two millennia of lead mining. Matthew wondered aloud what a Roman soldier must have thought when posted from the olive groves of Tuscany to the chilly hillsides of Somerset.

We were here to look for a very special bird, which we hoped would appear at dusk – roughly half-an-hour after sunset, which at this time of year is around 9.30pm. As we arrived, the sky was already showing those Turneresque shades of colour that are so hard to describe – and even harder to get a good photograph of. A song thrush was belting out its song from the conifer plantation across the road – do they ever stop? And as we waited, we had good views of a male stonechat and, even better, a slender tree pipit perched in a nearby tree with a beakful of insects for its brood of chicks, hidden somewhere nearby.

Then, just as the clock reached 10pm, the bird we had hoped to see appeared: a fast-flying, long-winged apparition with strangely jerky movements, rapidly heading away into the distance. For a moment or two it was possible to see its brownish, leaf-litter plumage and the two distinctive white flashes towards each end of its wings. It was, of course, a nightjar: a male, heading out at the start of the night to hunt for moths and other flying insects, which it would seize with its broad, open bill.

The sighting was a timely one, for the past week has seen the launch of 'Nightjar Nights', an online project celebrating this special bird, to which I contributed a short reading from one of my previous books. It felt fitting to go out and see these incredible birds for ourselves. We waited another half-an-hour or so, and were rewarded with a hooting male tawny owl, brief views of a hunting barn owl, and a second nightjar – a female, lacking her mate's white wing-flashes – which hovered like a giant moth over the long grass, before melting into the darkness.

Week Thirteen

15th–21st June

The End of the Beginning

Harold Wilson famously remarked that 'a week is a long time in politics'. He was wrong. Nowadays, with the advent of rolling news, social media and minute-by-minute updates on our smartphones, an hour can seem an eternity.

In the past week or two, we have witnessed an explosion of righteous rage, triggered by the unlawful killing of George Floyd, followed by peaceful Black Lives Matter demonstrations, the tearing down of a slaver's statue in Bristol, violent clashes between racist demonstrators and the police in London, and government ministers lining up to thwart the seemingly unthinkable suggestion by a twenty-two-year-old black man that the nation's children should not be the victim of institutional starvation. This was swiftly followed by a complete U-turn on providing free food for schoolchildren during the long summer holidays, led by a Prime Minister who appears to be doing the job for a bet, having claimed that he did not even know of footballer Marcus Rashford's open letter to MPs until fully twenty-four hours after the rest of us.

It's getting increasingly difficult to follow the twists and turns of the news, at a time when the nation – the entire world, for that matter – is going through the greatest existential crisis in history. And no, I'm not referring to Covid-19, but an even greater problem: the global climate crisis. As temperatures inside the Arctic Circle in Siberia rise to heatwave conditions – at a time when they would usually struggle to get into single figures – we are warned by an eminent scientist that the world has just six months in which to change the course of climate change and stop a post-lockdown rebound in greenhouse gases, which would destroy efforts to prevent a permanent climatic catastrophe.

At a time when even Brexit has been relegated to the inside pages of newspapers, and way down the TV and radio news bulletins, the chances of this dire and terrifying warning making any kind of impact seem to be, as a friend of mine bluntly put it, 'lower than whale shit'.

As I write, staring out of my office window into a uniform, grey drizzle, it seems hard to be optimistic about anything. Yet that same social media that raises my blood pressure to dangerous levels also brings me hope. Amongst the people I follow, and those who follow me, are a cohort of young people concerned with the environment, keen on hands-on conservation, but above all, enchanted by the natural world.

Almost a decade ago, when I wrote the *Natural Childhood* report for the National Trust, my conclusions on the future engagement of young people with nature were pretty gloomy. Today, I am delighted to admit that I was wrong to be so pessimistic: young people have seized the moment and are making real changes, both in the way we regard our relationship with the natural world, and by coming up with solutions to try to make it better.

I've already mentioned some of them, including six-teen-year-old Dara McAnulty, whose book *Diary of a Young Naturalist* has made such an impact. But there are so many more: Lucy McRobert, who the same year as I wrote that gloomy report founded the youth movement A Focus on Nature; Sorrel Lyall, who describes herself as 'a young, female, non-white birder', and who recently wrote a thoughtful Twitter thread pointing up the difficulties faced by people of colour when out birding; and Benedict Macdonald, whose book *Rebirding* is a practical manifesto for bringing back birds and other wildlife to one of the least biodiverse countries in Europe – the UK.

This generation has suffered more than most from lockdown: losing their jobs, unable to take their exams, or go to school, college or university, and being cooped up at home without their friends. Yet because they grew up with social media, they have embraced the lockdown better than anyone, posting uplifting stories, photos and videos of the world of nature on their doorstep. They have been at the forefront of a new engagement with the wild world that really has taken off during this crisis: not just amongst those already converted to the benefits of nature, like them and me, but through the rest of the population, who for the first time in their lives have been able to press the pause button on their busy lives and take in the world around them.

Most of all, these young people have done so while still making my generation – who are not just old enough to be their parents, but in some cases their grandparents – feel included in this new and exciting movement to change the world. We, in turn, have offered guidance and advice where we can, so that they perhaps avoid making the same mistakes as we did. The jury may still be out on whether or not this new generation will be able to save the world from Armageddon, but one thing is for sure: they are going to succeed or die trying. I salute all of them, from the bottom of my heart.

*   *   *

Meanwhile, things have been changing in our own small domestic sphere. Next week, George and Daisy go back to school – though for just one day a week until the summer holidays start. Our friend Mark left us a week or so ago, returning to his home in West London to continue his slow but steady recovery from the virus – though he's coming back this weekend to celebrate his birthday with us. And in a month or so, David and Kate will also depart, having planned to cycle from Land's End to John O'Groats in August. We will miss them more than we can say, as we also miss James, so far away from us in Japan.

Week Thirteen

We've booked a family holiday, in south Devon, for the end of August, when we hope and pray that most places will be open. And in the new spirit of 'seizing the day', inspired by David and Kate's plans, I dug out a decade-old plan to cycle the coast-to-coast route from the Irish Sea to the North Sea via Hadrian's Wall, called my old schoolfriend Simon, and booked the trip. Now all I have to do is get fit.

Lots of plans are, of course, still on hold. The chances of us flying to Australia this Christmas to see Suzanne's sister and her family are unlikely, to say the least; likewise, I am not sure when we will see James face-to-face again, after our spring trip to Japan was cancelled. I love travelling abroad to watch birds: but the twin disincentives of the Covid-19 crisis and the effects of long-haul air travel on climate change have put those plans on hold, perhaps indefinitely. I had better get used to the subtler joys of birding within a mile or so of my home.

What have been the advantages of lockdown? I hesitate to write about these, as I am painfully aware that for many millions of people the crisis has led to them losing their jobs, facing terrible financial hardship, being confined against their will to their homes, falling ill and in some cases either dying or losing a loved one, way ahead of time.

Yet I cannot pretend that I haven't found the whole experience mostly positive. Having to press the pause button – even for a few weeks – has made me rethink my life, in a good way. Being at home with our extended family, having time to relax and enjoy the small things in life as well as work, and above all, perhaps, rediscovering the joys of wildlife-watching close to home, have all been real gains.

Another thing I have been doing – along, I suspect, with many of my fellow-citizens – is watching a lot of TV. But rather than just flopping down on the sofa in front of whatever's on, I have been immersing myself in a mixture of comfort viewing – notably the excellent *Repair Shop*, which seems to sum up what is good about human nature – and what I suppose you would call 'discomfort viewing': long and detailed documentary or documentary-drama series about key events in history.

As well as working my way through *From the Earth to the Moon* – the compelling dramatized story of the Apollo space project that was so central to my 1960s childhood – I have also watched all ten episodes of Ken Burns' incredible documentary series *The Vietnam War*. It is tough viewing – yet also a parable for our own times, when so many governments and leaders, including our own, choose confrontation and division over cooperation and healing.

At the end of the final episode, Vietnam veteran, anti-war protestor and poet John Musgrave said something that resonated with me as much as anything else I have heard or read during the current crisis:

Being a citizen, I have certain responsibilities, and the largest of those responsibilities is standing up to your government and saying 'no' when it's doing something that you think is not in the nation's best interest. That is the most important job that every citizen has.

If one thing has changed during lockdown, it is this: we will no longer acquiesce in the way our country is being run – or misrun. As Peter Finch memorably put it in the

1976 movie *Network*, for which he posthumously won a best actor Oscar: 'I'm as mad as hell, and I'm not going to take this anymore!' That basically sums up how I feel.

But, after the crisis is over, and we return to 'normal' – or perhaps the better phrase is 'a new normal' – will things really change? I genuinely believe they will. Today, a National Trust survey revealed that people across the UK value nature more as a result of the lockdown. Specifically, one-third of all adults have become more interested in nature; more than half plan to spend more time outdoors; and over two-thirds have discovered that spending time in nature has made them feel happier. Best of all, the biggest growth in interest has been amongst people between the ages of twenty-five and thirty-four, traditionally the most disengaged cohort.

There has been a lot of talk about the 'wartime spirit' in the way we have responded, as a society, in our communities, as families and as individuals, to lockdown. On the day that the death of Dame Vera Lynn was announced – one of our last living links with the spirit of the Second World War – maybe we can at last shed the dangerous nostalgia for that conflict which taints so much of our government policy (notably Brexit) and recapture the real spirit of the war, and the post-war years: people standing in solidarity against a common foe, and working together for the collective good.

Epilogue

The Summer Solstice

When my children were younger, we would spend Father's Day looking for butterflies: specifically, rare butterflies that I had not yet managed to see. Back in 2007, we tramped up Collard Hill, on the scarp slope of the Poldens near Glastonbury, to search for – and find – the large blue. Two years later, on a breezy June day, we took a hike across Exmoor, where in a steep-sided combe we eventually found another target species, the heath fritillary.

Today, the day after the summer solstice, we are on a more leisurely family walk, around the loop with Rosie. There is a neat symmetry to our excursion; for here is where this book began, just three months ago.

As ever, Rosie runs a few yards ahead of us, sniffs the grassy verge, then turns around to make sure we are still there. The fields are bleached a straw-colour now, a far cry from the verdant green at the start of lockdown. The soundtrack is very different, too: just a few birds are still singing, including twittering goldfinches and newly fledged swallows, and a single whitethroat; but now the bleating of sheep predominates, as lambs and ewes call constantly to one another. The wind, too, is high in the mix, as it rushes through the reeds, some of which have grown so tall they almost block the view.

Spring is finally over, the nights are beginning to draw in, and the world has changed. It's not that nature disappears for the remaining half of the year; but it is harder to find, and the visual and aural onrush that marked the first few weeks of lockdown is now over – until next spring at least. I'll carry on walking around the loop with Rosie, and look forward to the subtler delights of late summer, autumn and then winter, as the seasons merge gradually into one another, like water rolling down a hill.

And then, just as the wind drops for a moment, we hear it: a brief snatch of the song of a skylark, somewhere in the sky above. Skylarks with Rosie, once again.

Postscript

The Winter Solstice

Just after eight o'clock, the sun begins to rise, marking the winter solstice.

On this, the shortest day of the year, the northern hemisphere turns away from the dark night of winter, and starts to lean back towards the light. From now on, little by little, the days begin to get longer, moving inexorably towards the Spring Equinox, when day and night are momentarily equal once again.

Meteorologically, the worst of winter is yet to come; we can, even at this time of climate crisis, expect frosts, snow and ice. Yet the lighter mornings and evenings will surely give us a new sense of hope, even at a time when the nation has been plunged into this nightmare of chaos and disarray.

Having lived through this tumultuous year, I don't believe that any of us will ever be able to think about the natural world in quite the same way again. Nature has always brought comfort and solace, but during the unique spring of 2020, it brought something more: a realisation that whatever we do, the world will carry on without us. Wild creatures were no more aware of the crisis than they are of any of our other worries and fears; this is a lesson we would do well to heed.

Another important lesson is that when we mess with nature, then nature will inevitably mess with us back. The longstanding idea that the natural world is there for us to exploit ad infinitum is no longer valid; if, indeed, it ever were. Now, at barely five minutes to midnight, we must permanently reset the way we live – not just for nature's benefit, but for ours, too.

So, despite all the continuing horrors, we should give thanks for one small but significant aspect of the spring lockdown. Paradoxically, by forcing us to stay local, in and around our homes, it allowed us to truly appreciate the global dimension of nature, as never before. Now, looking forward, we must build on this, to create a new and more equal relationship with the natural world.

Only if we do so will humankind – and the rest of nature – survive.

STEPHEN MOSS
Mark, Somerset
21st December 2020

Acknowledgements

At a time when my contact with others was, by circumstances, very limited, the list of people involved in this book is necessarily brief. Once again, I am indebted to my agent – and dear friend – Broo Doherty, who as ever brought her wise counsel and determination to ensure that it would be published.

This is the first time I have worked with Saraband, an independent publisher I have long admired. Sara Hunt, the founder and guiding light of the company, has been enthusiastic from the start; as has my editor Craig Hillsley, whose suggestions have improved the book enormously. My old friend Carry Akroyd has produced a wonderful cover, which really captures the time and place at the heart of the book. Huge thanks, too, to Rachael Bentley, who did the initial design. And another Saraband author, the peerless Donald S. Murray, kindly granted permission to use two of his wonderful lockdown poems, one as the book's epigraph. Thanks, too, to Rosie Hilton and Ruth Killick.

In the text itself, I acknowledge a number of people, ranging from old friends to people I have never met, whose contributions to social media entertained and inspired me during the lockdown period.

I would also like to thank all my colleagues and students, past and present, on the MA Travel and Nature Writing course at Bath Spa University. Over the past five years, they have constantly surprised me with the quality of their writing, which has in turn inspired me to improve my own.

Finally, a special thanks to the people who shared lockdown with me: Suzanne, Charlie, George, Daisy, David, Kate and Mark, and of course our fox-red labrador Rosie. Thank you for making this unique period in our lives so very memorable.